NUCLEAR MATERIALS AND DISASTER RESEARCH

DEPLETED URANIUM INDUCED PETKAU EFFECT

CHALLENGES FOR THE FUTURE

NUCLEAR MATERIALS AND DISASTER RESEARCH

Additional books in this series can be found on Nova's website under the Series tab.

Additional eBooks in this series can be found on Nova's website under the eBooks tab.

NUCLEAR MATERIALS AND DISASTER RESEARCH

DEPLETED URANIUM INDUCED PETKAU EFFECT

CHALLENGES FOR THE FUTURE

SVETLANA ZUNIC
AND
LJUBISA RAKIC

NOVINKA

New York

NOTICE TO THE READER

Library of Congress Cataloging-in-Publication Data

Names: Zunic, Svetlana, editor. | Rakic, Ljubisa, editor.
Title: Depleted uranium induced Petkau effect: challenges for the future / editor, Svetlana Zunic and Ljubisa Rakic (Clinical Center of Serbia and Serbian Academy of Sciences and Arts, Belgrade, Serbia).
Description: Hauppauge, New York: Nova Science Publisher's, Inc., [2016] | Series: Nuclear materials and disaster research | Includes index.
Identifiers: LCCN 2016021213 (print) | LCCN 2016027150 (ebook) | ISBN 9781634840460 (softcover) | ISBN 9781634853767 (ebook) | ISBN 9781634853767 ()
Subjects: LCSH: Depleted uranium-Health aspects. | Depleted uranium-Environmental aspects.
Classification: LCC RA1231.U7 D474 2016 (print) | LCC RA1231.U7 (ebook) | DDC 363.17/99-dc23
LC record available at https://lccn.loc.gov/2016021213

Published by Nova Science Publishers, Inc. † New York

CONTENTS

INTRODUCTION

As a naturally occurring, ubiquitous heavy metal, uranium is incorporated in the Earth's crust and it is a part of all natural resources. Natural and man-made products, including depleted uranium, differ in their isotopic composition, but all of them are α, β, and γ emitters, with a dominant alpha radiation emitted during their radioactive decay. Uranium remains radioactive for more than 4 billion years, exhibiting heavy metal and radiotoxic effects by an emission of alpha, beta, and gamma radiation. Depleted uranium has a unique potential to pose a threat to all natural resources including human society because of its radiotoxicity and its fissile properties including Thorium and Plutonium as well.

Even though the main source of exposure to ionizing radiation in the general population remains natural radiation, the exposure from medical procedures shows the increasing trend. Due to the uncontrolled military use of high amounts (a thousand tons) of depleted uranium, numerous unusual environmental physical manifestations were recorded in the last two or three decades. Simultaneous monitoring of natural phenomena on Earth and in the atmosphere has revealed an exceptional parallelism between the phenomena in the environment and in the living world. Our knowledge has evolved from *in vitro* studies of radiation exposure to a more comprehensive understanding of unexpected and poorly understood natural phenomena, whose consequences may be achievable according to the theory of litosphere-atomsphere-ionospehere and biosphere coupling.

The Petkau effect has been observed since 1972 as an inverse dose-rate effect *in vitro*. This study proposed possible mechanisms leading to the Petkau effect *in vivo*, taking into account the overall body integrative system's regulation. The health effects of depleted uranium on humans were observed in the functional unity of the lithosphere, atmosphere, ionosphere and biosphere. The Petkau effect was proposed as a wave phenomenon which can target all living structures, including cell membrane system or DNA molecules. On the other hand, one of the basic characteristics of living matter is its adaptability to the incident influence, up to the moment when the influence is powered enough to change the basic properties of a target, inducing its change to a different quality.

Genetic susceptibility is one of the key connecting links between adaptability and tissue damage with the possible evolution of the neoplastic change. We discussed the importance of an individual approach to the diagnosis and selection of appropriate therapy, based not only on the results of the expression analysis, but also on the metabolic and apoptotic tissue properties.

Keywords: depleted uranium radiobiology, Petkau effect, alpha-particle-induced bystander effect, Gulf War Syndrome, Balkan Syndrome

Svetlana Zunic
Clinical Center of Serbia, Belgrade, Serbia
Email: svetlana.zunic@gmail.com

Ljubisa Rakic
Serbian Academy of Sciences and Arts, Belgrade, Serbia.

ABBREVIATIONS

AC	apoptotic capacity
AI	apoptotic index
AM(s)	alveolar macrophage(s)
ATP	adenosine triphosphate
BAL	bronchoalveolar lavage
BSE	bystander effect
DU	depleted uranium
ELF	extremely low frequency
EMF	electromagnetic field
EMR	electromagnetic radiation
EU	enriched uranium
HSP(s)	heat shock protein(s)
LEC	Lupus Erythematosus cells
LET	linear energy transfer
MN	micronuclei
NSCLC	non-small-cell-lung cancer
NU	natural uranium
ROS	Reactive oxygen species

Chapter 1

DEPLETED URANIUM RADIOCHEMICAL PROPERTIES

Uranium is a naturally occurring, ubiquitous, heavy metal. Natural uranium (NU), in various chemical forms, is found in all soils, rocks, seas and oceans. The Earth's crust contains an average of about 3 ppm (= 3 g/t) uranium, and seawater approximately 3 ppb (= 3 mg/t) (WISE, 2011). Uranium is also present in drinking water and food. Natural uranium consists of a mixture of three different isotopes: U-238 (99.27% by mass), U- 235 (0.72%) and U-234 (0.0054%) (WHO, 2001). U-235 is a fissile radioactive isotope, which is a primordial nuclide, existing in nature in its present form since before the creation of Earth (WISE, 2011).

From normal intakes of water, food and air, about 90 µg of uranium exist in the human body (Priest, 1990). The skeleton contains about 66%, the liver 16%, the kidneys 8% and the other tissues approximately 10% (WHO, 2001).

Uranium is used primarily in nuclear power plants; most reactors require uranium in which the U-235 content is enriched from 0.72 to about 3–4% in enriched uranium (EU). The uranium remaining after removal of the enriched fraction is referred to as depleted uranium (DU) (WHO, 2001-1). The waste product from the enrichment process is depleted in uranium-235. The concentrations of U-235 in DU are 0.2 to 0.3 weight-percent; that is around 30 - 40% of its concentration in NU (WISE, 2011). The concentration of uranium-234 is depleted to an even lower ratio, according to its lower atomic weight. DU is weakly radioactive and a radiation dose from it would be about 60% of that from purified natural uranium with the same mass (WHO, 2003).

Table 1 presents the properties and composition of natural, enriched and depleted uranium (according to WISE, 2011).

Table 1. Properties and isotopic composition of natural, enriched and depleted uranium

	U-234	**U-235**	**U-238**
Half-life (years)	244,500	$703.8 \cdot 10^6$	$4.468 \cdot 10^9$
Isotopic composition (weight %)			
NU	0.0053%	0.711%	99.284%
DU	0.0008976%	0.2%	99.799%
EU	0.02884%	3.5%	96.471%
Isotopic composition (activity %)			
NU	48.9%	2.2%	48.9%
DU	14.2%	1.1%	84.7%
EU	81.8%	3.4%	14.7%

U-238 and U-235 are the parent nuclides of two independent decay series, while U-234 is a decay product of the U-238 series. Tables 2 and 3 present the U-238 and U-235 decay series, assuming that all decays emit gamma radiation (according to Clark, 1996 and WISE, 2011). U-238 has a half-life of 4.5×10^9 years, which is approximately, the age of the Earth (Barnes, 1996). The U-238 gamma spectrum is not in equilibrium with U-234 and its daughters. U-235 relatively quickly decays to equilibrium with its daughters, Th-231 and Pa-231 (Clark, 1996). One of the daughters of U-238 is Rn-222 (Table 2). Two daughters of Rn-222, Po-218 and Po-214, are alpha emitters that can adhere to the respiratory tract (Barnes, 1996).

Natural and depleted uranium differ in their isotopic composition (Table 1), but both are α, β and γ emitters, with a dominant alpha radiation emitted during their radioactive decay (Tables 2-3).

DU has less of the more radioactive isotopes uranium-234 and uranium-235 per unit weight than natural uranium (Burger, 2012). The International Atomic Energy Agency (IAEA) defines DU as a low specific activity material. The specific activity of uranium in DU is about 15 Bq per mg compared with 25.4 Bq per mg for natural uranium. If the activity of the decay products is also included, then the value for specific activity is higher. Plutonium (for example, plutonium-239) has a much higher specific activity of about 2,300,000 Bq per mg (PHE, 2007).

Table 2. Uranium-238 Decay Chain

Nuclide	Half-Life	Radiation*
U-238	$4.468 \cdot 10^9$ years	α (γ)
Th-234	24.1 days	β (γ)
Pa-234m	1.17 minutes	β (γ)
U-234	244,500 years	α (γ)
Th-230	77,000 years	α (γ)
Ra-226	1,600 years	α (γ)
Rn-222	3.8235 days	α
Po-218	3.05 minutes	α
Pb-214	26.8 minutes	β (γ)
Bi-214	19.9 minutes	β (γ)
Po-214	63.7 microseconds	α
Pb-210	22.26 years	β (γ)
Bi-210	5.013 days	β
Po-210	138.378 days	α
Pb-206	Stable	-

* Almost all are γ-emitters.

Table 3. Uranium-235 Decay Chain

Nuclide	Half-Life	Radiation *
U-235	$703.8 \cdot 10^6$ years	α (γ)
Th-231	25.52 hours	β (γ)
Pa-231	32,760 years	α (γ)
Ac-227	21.773 years	β
Th-227	18.718 days	α
Ra-223	11.434 days	α
Rn-219	3.96 seconds	α
Po-215	778 microseconds	α
Pb-211	36.1 minutes	β
Bi-211	2.13 minutes	α
Tl-207	4.77 minutes	β
Pb-207	Stable	-

* Almost all are γ-emitters.

Depleted Uranium Is a Radiotoxic Heavy Metal

DU exerts mixed, radioactive (α, β, γ emitter's) and toxic heavy metal properties. Our knowledge concerning uranium or DU toxicity has evolved since 1999, when DU was considered as a Group III agent (not classifiable as carcinogenic to humans) by the International Agency for Research on Cancer (IARC). According to Baverstock (2006), DU has been categorized as a Group I agent – alpha emitter (i.e., as carcinogenic to humans).

Uranium decay produces alpha particles, which include two protons and two neutrons, beta particles, which are essentially high-speed electrons and gamma rays which are photons.

Heavy nuclei emit an alpha particle to reduce the number of protons. Typically, alpha decay leads to daughter nuclides that themselves are radioactive (usually, alpha or beta minus emitters). When an alpha particle (identical to a He nucleus without the orbiting electrons) is emitted from a nucleus, a new element is formed with the mass number decreased by 4, and the atomic number decreased by 2 (Barnes, 1996).

The alpha particle is ejected with a large, precisely defined energy usually between 4 and 10 MeV (Barnes, 1996). The alpha particle mass is 6.644656×10^{-27} kg. Most alpha particles have kinetic energies in the range of 3 to 7 MeV, or a typical kinetic energy of 5 MeV and the speed of 15,000 km/s (Tipler and Llewellyn, 2002; Benton, 2015).

Alpha particles have huge ionization potential because of their high energy. Alpha particles exhibit high ionization impact because of their relatively large mass, +2 charge, and relatively low velocity. Alpha particles lose their energy in interaction with other atoms, within a few centimeters of air, or travel only a distance of microns in soft tissue before stopping (Barnes, 1996). Even though alpha particles have the smallest penetrability, their biological effects are the most harmful.

If the ratio of neutrons to protons in the nucleus is too high, beta decay occurs. When a neutron is converted into a proton the radionuclide becomes a different element whose nucleus has one neutron less and one proton more (the atomic number Z increases by 1, while the mass number A remains the same), and an electron is ejected. A gamma photon can be emitted if the energy released from the nucleus is not enough, to stabilize the nucleus. Beta particles are high-energy, high-speed electrons ("beta minus" (β^-), or positrons "beta plus" (β^+)) (Jevremovic, 2009).

Gamma rays are released during uranium decay to provide more stability to the nuclei after α or β particles emission. Gamma rays are high-energy photons that travel at the speed of light along hundreds and thousands of meters, easily penetrating and passing through matter, including human tissues. Gamma rays have no mass and no electrical charge; they may be stopped by very dense materials, such as lead. Gamma radiation is an electromagnetic radiation with the shortest wavelengths and highest frequencies. When passing through matter, gamma radiation ionizes atoms directly through the photoelectric effect and Compton scattering, and indirectly through pair production (Barnes, 1996).

The energy of the gamma-ray of pure U-238 is 49.55 keV (Clark, 1996). The alpha decay of U-235 to Th-231 is accompanied by the emission of a prominent gamma ray at 185.7 keV (4.3 x 10^4 of these 185.7 keV gamma rays are emitted per second per gram of U-235). The relatively low energy and consequent low penetrating power of these gamma rays implys that most of those emitted within the interior of the material are absorbed within the material itself (U.S. Atomic Energy Commission, 1974). As for the photon energies which are used in nuclear medicine (50 keV – 5 MeV), there are three types of interactions of photons with matter we are interested in: 1) In Compton scattering, a photon interacts with an outer-shell electron in an atom of the material that the photon is traversing. The photon scatters from the electron through a scattering angle. The electron is set into motion with energy equal to that which the scattered photon loses (Barnes, 1996). Compton effect is the principal absorption mechanism for γ-rays in the intermediate energy ranges (100 keV – 10 MeV) (Gupta, 2013); 2) Photoelectric absorption of a photon by an atom results in the ejection of an inner-shell electron. The energy of the ejected electron (photoelectron) is the same as the initial photon energy, minus the binding energy of the electron (Barnes, 1996). The photoelectric effect is the dominant energy transfer mechanism for γ-rays with the photon energies below 50 keV (Gupta, 2013); 3) In the process of pair production, the energy of the photon is converted into mass, an electron and positron pair (Barnes, 1996).

Depleted uranium is very dense (19,050 kg/m^3). It is denser than other known substances under standard (i.e., Earth-surface) pressures, like lead, osmium or iridium (DOE-HDBK, 1994; WISE, 2011). DU is an alternative for tungsten, which is more expensive and has fewer offensive capabilities.

Due to its high density, about twice that of lead, DU has several civil applications: in the coloring of ceramics and glass, development of dentistry technology, as a chemical catalyst, for radiation shields in medical equipment, or as containers for the transport of radioactive material.

The military use of DU is in munitions designed to penetrate armor plates, and it is also used to reinforce military vehicles (Betti, 2003; WHO, 2003). The munitions containing DU were used in Iraq during the 1990–1991 and 2003 Gulf wars, in Bosnia and Herzegovina in 1994–1995, and in Serbia and Montenegro in 1999 (Burger, 2012).

Although some measurements of DU at sites where DU munitions have been used indicate only localized contamination at the ground surface (Keller et al., 2001), the extent and type of contamination may imply that significant quantities of depleted uranium entered the water supply and food chain and ground water (WHO, 2001).

Depleted Uranium Exposure Routes

Depleted uranium has been *repeatedly* used by the military, approximately every four years since 1991 (Iraq 1991, Bosnia 1994-1995, Kosovo, Serbia, and Montenegro 1999, Afghanistan 2001-2003, Iraq 2003-2011), which has induced the *low dose* radiation (air pollution easily transferable to the remote distances from the place of explosion), *slow doses* (the DU ammunition remnants can be fully oxidized into corrosion products twenty-five to thirty-five years after impact) (Burger, 2012) and its further prolonged contribution to the maintenance of alpha particles radiation (Zunic and Rakic, 2013).

Civilians have limited access to information concerning the exact amount of missiles used, the number of fragments and unexploded ordnance, or the people exposed, as well as the exact contamination levels. Depending on aerosol speciation, inhalation may lead to a protracted exposure of the lung and other organs (Bleise, 2003). One 120 mm DU tank round impacting against a hard target will create about 950 grams of DU dust, while one burst of 30 mm shells, fired by an A-10 aircraft, might create 960 grams. In some cases, the amount of dust created may be much higher. When a DU shell hits a hard target DU burns. About 20% of DU vaporizes into a fine dust that can travel long distances, and be inhaled by people in the immediate vicinity (up to 400 meters) or up to thousands of miles away (Military Toxics Project Information Sheet, 2003).

The firing of DU munitions can immediately contaminate air, soil, and water with the ingestible radio-chemotoxic DU particles (Military Toxics Project Information Sheet, 2003). Gatti and Montanari (2004) reported that the combustion processes created a form of particulate pollution that could be released into the environment. The size of the particles inversely relates to the temperature of the process. In case of DU, the temperature is higher than 3,000 °C. Consequently, inorganic micro- and nano-particles have been generated, which could pollute the environment after the explosion of DU missiles. There are three major uranium oxides produced by burning. These are UO_3, U_3O_8, known as uranium trioxide, triuranium octoxide and uranium dioxide, UO_2. Although uranium is one of the densest metals known, the uranium oxides in the smoke and dust are not so dense and remain suspended in the air for a long time. These aerosols can contain very small particles of uranium oxide of between 0.1 and 10 microns in diameter, which can be inhaled and deposed in the lungs (Al-Muqdadi and Al-Ansari, 2009).

Unless shells and fragments are removed from the areas of use, they will continue to release DU into the environment for years or decades. The corrosion of the spent shells adds to the amounts of the mobile DU dust in the environment (Military Toxics Project Information Sheet, 2003). The depleted uranium oxide dust, produced when DU munitions burn, is composed of two oxides: one insoluble and the other sparingly soluble. Their particle size and distribution is not typical as for particles normally encountered in radiological protection (Baverstock, 2006).

Busby and Morgan (2006) tried to answer the question whether the use of uranium weapons in the Second Gulf War resulted in contamination in Europe. The authors found an excess of uranium in the air along the trajectories across Europe, of some 500 nBq/m^3, assuming that uranium particles originated from the Persian Gulf battlefields. It was found about 48,000 particles of 0.25 μm diameter in one cubic meter. By the authors' approximate estimation, each person would have inhaled about 23 million particles of uranium in six weeks. One half of the total mass of the uranium oxide consists of particles smaller than the wavelength of visible light. The behavior of DU particles may be taken approximately to that of a gas, whose dispersion may be expected to be similar to the dispersion of radioactive gases from nuclear accidents like the Chernobyl accident (Busby and Morgan, 2006).

The meteorological conditions at the time of the initial bombing in the Second Gulf War were anomalous, with airflow from Iraq to Europe. During February and in April 2003, this airflow carried Saharan desert sand to the UK (Busby and Morgan, 2006). The DU air concentrations in European countries closer to Iraq would have been exposed to higher levels of radioactive particles, contributing to significant exposure to the public (Busby and Morgan, 2006). The annual quantity of desert dust that makes regional or global airborne migrations is 0.5 to 5.0 billion tons (Perkins, 2001). Sandstorms and fires were recorded by NASA on the images of Iraq, Kuwait, and parts of Saudi Arabia and Iran that were acquired on March 20&21, 2003 (the Moderate Resolution Imaging Spectroradiometer (MODIS) on the Terra satellite late morning local time in Baghdad) (NASA, 2003).

The spreading of radioactive dust originating from DU from battlefields, across remote regions around the globe has been assisted by natural phenomena, including sandstorms. The larger deserts on the planet, which include the Sahara and Sahel regions of North Africa and the Gobi, Takla Makan, and Badain Jaran deserts of Asia, are the primary sources of mobilized desert topsoils that move great distances through the atmosphere each year. During summer in the Northern Hemisphere, African desert dust is transported across the Atlantic to the northern Caribbean and North America. During winter in the Northern Hemisphere, African desert dust is transported across the Atlantic to the southern Caribbean and South America. The Asian dust season typically lasts from late February to late April. Large Asian dust events can travel significant distances in the Northern Hemisphere (Griffin, 2007).

DU armaments have repeatedly been used during the last 25 years: in Iraq (1991), Afghanistan (2001-2003), Iraq (2003), in numerous conflicts in North Africa or West Asia, which are just the regions where dust storms usually occur. As well as infective agents (Griffin, 2007), micro DU particles may be transferred along sand to distant regions of the world.

Thus, the affected territory spreads around the globe. Airflow from Iraq to Europe carried Saharan desert sand way to the UK (Busby and Morgan, 2006). The area which is potentially exposed to contamination with DU particles, based on reports of Busby and Morgan (2006), Moret (2006), Schwarz (2006), Zunic (2013; 2013-1) is much wider, including ultimately, the whole world. According to Zunic and Rakic (2015), the potentially exposed area after the bombing of targets in the Persian Gulf and the Balkans and Afghanistan with depleted uranium projectiles is shown in Figure 1.

Uncertainty about the DU particles' behavior in the environment, or after entering the body, made an incomplete risk assessment for exposure to DU used in military campaigns (RS, 2001; RS, 2002). The impact of DU on the environment provoked some new attitudes in an understanding of the complex and unpredictable mechanisms of interaction of DU with all natural resources.

Figure 1. Schematic presentation of the potentially exposed area after the bombing of targets in the Persian Gulf, the Balkans and Afghanistan with depleted uranium projectiles. The thick black circle outlines approximately 2,400 miles radius around the Persian Gulf. The thin black circle schematically represents approximately 2,400 miles around Bosnia and Herzegovina. The dashed black circle schematically represents approximately 2,400 miles around Serbia. The punctuated circle schematically represents approximately 2,400 miles around Afghanistan. Adapted from primary data of Zunic, (2013).

DU particles undergo alpha, beta, and gamma decay, which is achievable and inevitable. Nevertheless, the interaction of radioactive particles of depleted uranium with molecules and ions in the environment, or with the constituents of living systems is unpredictable and insufficiently explored. In fact, depleted uranium has been declared as a "safe" weapon that has been widely used for a quarter of a century. As a consequence, we are unfortunately faced with unpredictable and unknown environmental phenomena, as well as rare, new and no recognizable biological effects and medical consequences.

Chapter 2

ENVIRONMENTAL EFFECTS OF DEPLETED URANIUM

Military uses of DU may have a significant impact on the environmental equilibrium of the uranium isotopes. Every change that is high enough to modify the ionic, magnetic, or temperature Earth's equilibriums, depends on the natural default globe properties and tends to reach this equilibrium again. It has been reported recently that a thousand tons of DU that have been used since its first military use in the Persian Gulf in 1991 to date, have changed sufficiently the Earth's natural equilibrium in terms of default activity of natural uranium in the Earth's crust and have triggered the visible output(s) (Zunic and Rakic, 2013). The authors have estimated that the output retains its value until the input changes sufficiently to trigger a change. Every new use of DU in military campaigns, or a discharge of radioactivity in civil nuclear disasters, may intensify the output.

The repeated use of depleted uranium can produce ionizing radiation that, above a certain (unexplored) threshold, may trigger the disproportionately high response to the level where it becomes unpredictable and gives empirically unknown consequences. When the input is below a hypothetical preset threshold (natural properties of Earth), the output is absent to low, and the records can be confusing, or misinterpreted (according to Stein, 2013). In the environment, radiation hormesis is feasible, sometimes with concomitant catastrophic natural phenomena (Zunic and Rakic, 2013).

The uncontrolled military use of high amounts of DU has coincided with numerous unusual environmental physical manifestations that have been recorded in the last 20 years. Simultaneous monitoring of natural phenomena on Earth and in the atmosphere has revealed an exceptional parallelism between the phenomena in the environment and in the living world: increased number of earthquakes (Ellsworth, 2013), elevated humidity in the environment (IGMASS, 2012), increased number of forest fires (Jovanovic and Oldja, 2007; Rekacewicz, 2007), and increased extreme weather events during the last 20 years (EEA, 2012).

The climate changes directed the focus of our thinking to the question whether periodical, artificial discharge of the large amounts of ionizing alpha particles emitted from the decay of DU that was used for military purposes, can seriously misbalance the nature equilibrium conditions. There is a remarkable parallelism between the use of DU ammunition and the physical phenomena described by the sources cited above. A number of reported extreme weather events and wildfire in EEA member and collaborating countries (1980–2011) (EM-DAT, 2012) showed that extreme weather events were mostly frequent in 1990, around 2000 and later, which was going on in parallel with the frequent use of DU in military actions since 1990. The reported extreme weather events were mostly frequent in the period 2000-2010, which coincided with the excessive use of DU for military purposes during the Second Gulf War.

These data support our hypothesis that after the local military conflicts during which DU ammunition was used, an unpredictably wide territory has been contaminated by aerosols, and later water and ground natural resources. From the air, the particles fall very slowly and contaminate the ground and grass, vegetables, fruit, entering the alimentary chain. From the rain, those particles could penetrate the earth and enter springs and subterranean waters (WHO, 2001).

More precisely, the entire territory of Europe was exposed to DU contamination at the time of military operations during which radioactive ammunition was used. Numerous other studies support this claim. Forest fires in the Balkans were the most frequent exactly during the bombing of targets in Serbia and Bosnia (Figures 2 and 3).

Rekacewicz's graph (UNEP/GRID-Arendal, 2007) highlighted the gradual increase of the frequency of forest fires over the past 20 years in Southeast Europe (Figure 2).

The prominent peaks were detected in the period 1995-1997 (after the bombing of Bosnia and Herzegovina, which borders Serbia), and 1999-2001, after the bombing of Serbia in 1999 (Figure 3).

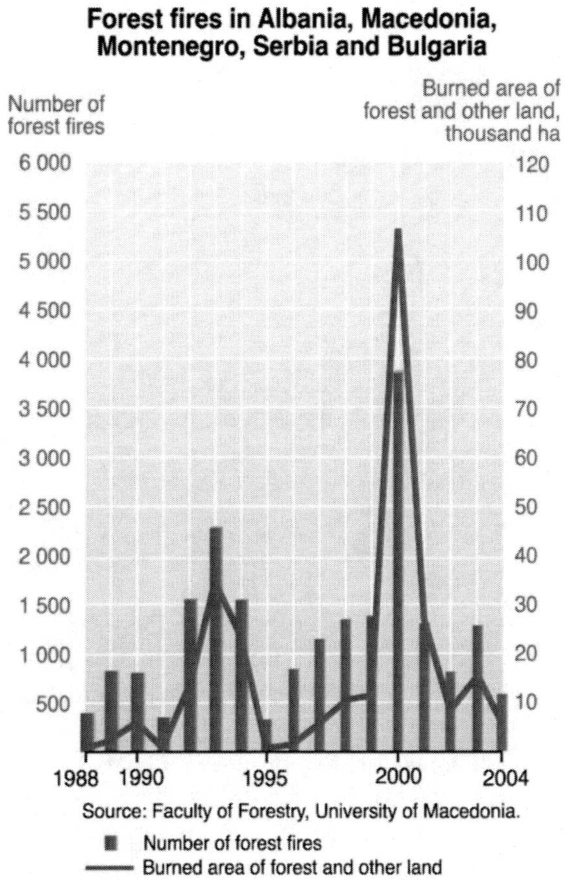

Figure 2. The number of forest fires in Albania, Macedonia, Montenegro, Serbia and Bulgaria and their extensiveness (Reproduced with permission of Rekacewicz, P. Forest fires in Albania, Macedonia, Montenegro, Serbia and Bulgaria. [Balkan Vital Graphics]. *Environment without borders UNEP/GRID-Arendal* [2007]. (Available online at http://www.grida.no/files/publications/balkan-vital-graphics/balkans-vital-graphic-full.pdf).

The comparison of Figures 2 and 3 showed a remarkable correlation between the forest fires number and their extensiveness for the estimated period of time in Serbia and its neighboring territories. The peak corresponding to the year 2000 may be a result of forest fires sweeping across vast areas of the Republic of Serbia, caused by dispersed parts that originate from DU projectiles.

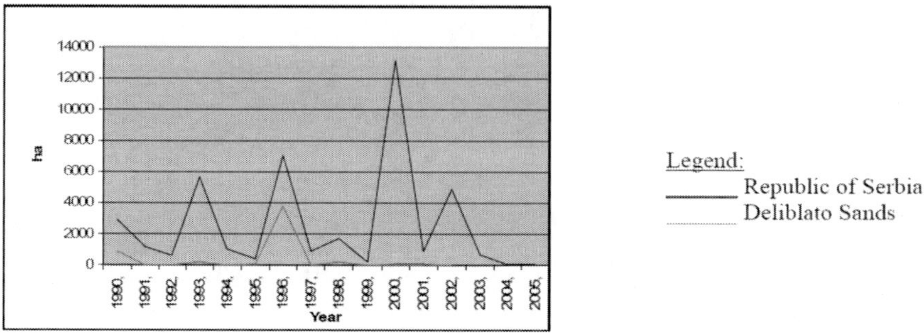

Figure 3. Review of the burned areas by year (1990-2005). (Reproduced with permission of Jovanovic, V; Oldja, M. [Video control of wildfires in Serbia. Ministry of Agriculture, Forestry and Water Management]. Directorate of Forests, Belgrade Serbia and Public Enterprise "Vojvodinasume. [2007]. Available online at http://www.fire.uni-freiburg.de/sevilla-2007/contributions/doc/SESIONES_TEMATICAS/ST7/Jovanovic-Oldja_SERBIA.pdf.

Uranium is pyrophoric (i.e., the reaction of the metal with oxygen in the air causes it to ignite spontaneously) (Burger, 2012). It is possible that forest fires may be caused by the friable remnants of exploded armaments with DU, or by unexploded items in that region. For example, from 1990 to 2005 there were more than 1,700 forest fires in the Republic of Serbia. The total burnt area of forests was about 40,000 ha (Jovanovic and Oldja, 2007).

Every flaming of contaminated land or plants, can re-suspend the absorbed radioactive particles and burn them up into the air. There is a prolonged and increasing risk of re-contamination of areas that were contaminated in peace time or war. Some of the materials that were contaminating that area would have been incorporated into the woods. In other words, they landed on the ground in 1986 and got absorbed into the trees and the entire biosphere (Busby and Morgan, 2006). When they burned, they just became re-suspended. "All of that material which fell on the ground will now be burned up into the air and will become available for people to breathe" (Zinets and Prentice, 2015).

DU and Climate Forcing

There are many natural climate forcings inducing changes to the Earth's climate system. Besides natural climate forcings, like large volcanic eruptions that inject light-reflecting particles as high as the stratosphere, man-made climate forcings relate mainly to particle pollution (aerosols) (Lindsey, 2009).

The repeated military use of depleted uranium nuclear weapons caused contamination both at the local and global scales, because of the movement of air masses from the location of uranium missiles explosion all around the Earth (Zunic, 2013; 2013-1; Zunic and Rakic, 2015). The firing of DU munitions can immediately contaminate air, soil, and water with ingestible toxic and radioactive DU particles. The corrosion of spent shells adds to the amounts of the mobile DU dust in the environment (Military Toxics Project Information Sheet, 2003).

A forcing can trigger feedbacks that intensify (positive feedback) or weaken (negative feedback) the original forcing. For example, loss of ice at the poles, which makes them less reflective for solar irradiation, is an example of a positive feedback (NASA, 2009). The atmosphere heating effects of military used DU could be better clarified by the fact that during the complete fission of 1 kg U-235, 19 billion kilocalories are released, i.e., 1 kg U-235 corresponds to 2.7 million kg coal equivalent (data available online at: https://www.euronuclear.org/info/encyclopedia/coalequivalent.htm). If about 3,000 tones of depleted uranium had been used in the military campaigns (1991-2011), then, according to Table 1, approximately 6,000 kg of U-235 have the equivalent heating effect as if about 16,200 million kg of coal have burnt at the Earth's surface!

This may be one of the reasons why during the 20th century and the last two decades, the global mean sea level rose at rates of 1.7 mm/yr and 3.2 mm/yr respectively, as a result of both increase of ocean thermal expansion and land ice loss. For the period 1993–2010, glaciers and ice caps have accounted for ~30% of sea level rise (Allison et al., 2009).

INTERACTION OF DU PARTICLES WITH SMOKE

Alpha particles from any source are easily stopped by smoke. The small particles of radioactive fumes originating from the battlefields, which comprise micro or nano-sized charged particles of depleted uranium, can be adsorbed on smoke and dust particles in the atmosphere. The radon concentrations measured in Iraq on some battlefields may originate from natural uranium (Keller et al., 2001), but radon may also come from the radioactive decay of DU that was used in the Persian Gulf military actions.

Radon mostly appears with the decay chain of the radium and uranium series, and marginally with the thorium series.

The excess of positively charged ionizing alpha particles that are deliberated from uranium decay can be temporarily concentrated by the air pollution above big cities. Ionization induced by these alpha particles may trigger there local changes of the electromagnetic properties in the atmosphere, and heating as well.

The big cities-big earthquakes relation has been outlined in one of the Science releases (Stein, 2013). Santiago and Tokyo were in the areas of increased seismic activity, but…the data given by Oyama (2012) on his website reported that alpha ray on Japan's seaside had been too high even compared to the contamination situation in Fukushima. He suspected that it might be from past nuclear tests.

We wonder if the seismic activity may be provoked there in regard of the Lithosphere-Atmosphere-Ionosphere Coupling (LAIC) Model (IGMASS, 2012; Zunic and Rakic, 2013).

LITHOSPHERE-ATMOSPHERE-IONOSPHERE AND BIOSPHERE COUPLING

Repeated abrupt release of ionizing particles in the atmosphere (military or peacetime disasters) may induce increased air ionization with a consequent increase in the atmosphere heating (Zunic and Rakic, 2013).

Alpha particles from natural sources, nuclear disasters in peace-time and after military use of several tons of depleted uranium, are emitted during uranium decay. A rapid increase of DU, and consequently positively charged α-particles in the atmosphere, may induce changes in the electromagnetic field and in the coupling mechanism in the lithosphere, atmosphere and ionosphere, as presented in Figure 4.

Traveling at approximately one-twentieth the speed of light, alpha particles strike air molecules and eject electrons. Positively charged ions and negatively charged electrons are the products of air ionization induced by alpha particles. Due to their electric charge and large mass, alpha particles lose energy rapidly along the path of only a few centimeters in the air. After losing the energy, alpha particles are converted to helium by binding of free electrons (Smith, 2010).

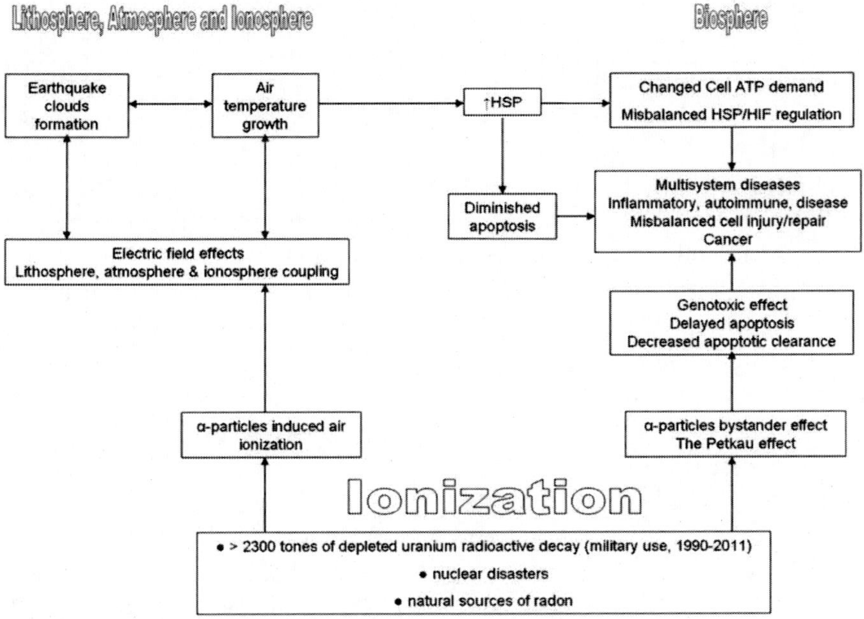

Figure 4. The impact of alpha particle induced ionization on the lithosphere-atmosphere-ionosphere and biosphere coupling. Adapted from primary data of Zunic and Rakic, (2013).

The Lithosphere-Atmosphere-Ionosphere Coupling (LAIC) Model, given by Sergey Pulinets, highlights the strong influence of radon decay generated alpha particles and its ionization onto the atmosphere heating, as well as their importance in the earthquake clouds formation. The atmosphere above Japan heated rapidly before an earthquake. Infrared emissions above the epicenter increased dramatically, which was in relation to the releases of large amounts of radon in the days before the devastating earthquake in Japan (MIT Technology Review, 2011).

Based on the simultaneous occurrence of physical phenomena and health effects that could be monitored during and after the artificial use of depleted uranium for military purposes in the Persian Gulf and the Balkans (Zunic, 2013; Zunic, 2013-1), conclusion is that it is impossible to distinguish the environmental effects in terms of the origin of alpha particles (from natural sources or man-made isotopes, including nuclear weapons).

DU Particles May Induce Light in the Atmosphere

Light emission induced by alpha particles in the air was first observed by Sir William and Lady Huggins in the early years of the 20th century (Sand et al., 2014).

It has already been highlighted that mobile air masses containing nano-sized particles of DU contaminate the atmosphere globally, all around the Earth (Zunic and Rakic, 2015). Given the Earth's overall diameter and the surface, the local areas of military campaigns where bombs containing depleted uranium have been used can be comprehended as the stippled sources of imprinting uranium dusts into air, after which α and β particles and gamma radiation are emitted during uranium radioactive decay.

The Earth can be regarded as a nearly conducting sphere, around which the thin, dielectric atmosphere extends up to the ionosphere. The atmospheric electric discharges generate broadband electromagnetic waves that propagate between the surface and the ionosphere (Simões, Pfaff and Freudenreich, 2011). Bearing in mind the proposed geometrical and atmospheric attenuation factors and adding thunderstorms roll over Earth, producing some 50 flashes of lightning every second, we have understood that each lightning burst creates electromagnetic waves that begin to circle around Earth. Some of the waves may combine to form the Schumann resonance. The Schumann Resonance is a standing wave (around 8 Hz) in the atmosphere (Miller, 2013).

The specific ionization of an *alpha particle* is very high, in the order of thousands of ion pairs per centimeter of air (Hoff, 2012).

Alpha particles induce ultraviolet (UV) air fluorescence, which is mainly emitted by molecular nitrogen, the most abundant gas in the air (Bachelor et al., 2009; Hoff, 2012). Alpha particles provide excitation energy to nitrogen and the gas emits fluorescence, or photons in the air. Each time this occurs, alpha particle loses kinetic energy (Hoff, 2012). Nitrogen emits fluorescence at wavelengths in the range of 300 to 400 nm. The nitrogen fluorescence induced by radiation can be used to detect the presence of ultra-high energy cosmic rays interacting with the Earth's atmosphere and for the detection of radioactive contamination of the environment (Bachelor et al., 2009). Humidity can affect the results, since water molecules, together with oxygen, are effective quenchers for excited nitrogen molecules (Waldenmaier 2006).

The speed of a *beta particle* depends on its energy. The energy of beta particles can vary over a wide range. When beta energy exceeds the thermal level, the breaking of chemical bonds occurs, resulting in ions formation. In the environment or in the body, after losing its energy, a beta particle, as an electron, interacts with a positive ion. The physics of electron-induced luminescence in the air is studied by many authors (Waldenmaier 2006; Lefeuvre et al., 2007; Abbasi et al., 2008). Due to differences in the mass and charge of the initial particles (alpha or beta), the excitation efficiency is different, but the luminescence is mostly induced by collisions with secondary electrons and nitrogen molecules in both cases. The luminescence induced by charged particles is mostly proportional to the energy loss in the media, and is not very sensitive to the type of primary particle (Sand et al., 2014).

The total energy deposited to air by low-energy beta particles (average energy 5.2 keV) is over 2,600 times smaller than the energy deposited by alpha particles. The contribution to the total light emission by β-particles from DU decay can be neglected. Luminescence is excited only after a significant number of electron–electron interactions (Sand et al., 2014).

The interaction of *gamma rays* with matter causes the generation of other charged particles such as positrons and electrons at relativistic speeds. Very short blasts of terrestrial gamma-ray flashes are recorded due to the fact that electrons travel at nearly the speed of light and scatter off of atoms decelerating in the Earth's upper atmosphere. The exact mechanism that accelerates the electron beams to produce terrestrial gamma photons is still uncertain, but it probably involves the build-up of electric charge at the tops of thunderclouds due to lightning discharges. The consequence is a powerful electric field between clouds' tops and the ionosphere (NASA, 2005).

Electrical Coupling of Lithosphere-Atmosphere-Ionosphere and Biosphere

A new model of generation of electric field on the basis of injection of the charged aerosols into the atmosphere has been discussed in order to provide better insight into the mechanism of lithosphere-atmosphere-ionosphere coupling (Sorokin and Hayakawa, 2013). The authors have found that seismic activity is accompanied by an injection of radioactive substances into the atmosphere, which leads to the changes of the state of the ionospheric plasma and electromagnetic field. These changes in the electromagnetic properties of the Earth's mantle could be detected a few days before an earthquake, as we have hypothesized in our recent publication (Zunic and Rakic, 2013).

We have also postulated a hypothesis on the lithosphere-atmosphere-ionosphere and biosphere coupling (Figure 4) (according to Zunic and Rakic,

2013). The uncontrolled military use of high amounts (over 2,300 tons) of DU, in parallel with the physical manifestations that have been recorded during the last decades, based on the simultaneous monitoring of natural phenomena on Earth and in the atmosphere, directed the focus of our thinking to the question whether periodical, artificial discharge of large amounts of ionizing alpha particles emitted from the decay of uranium, which was used for military purposes, can seriously misbalance the nature equilibrium conditions and human health (Zunic and Rakic, 2013).

Chmyrev and coworkers (2013) discussed the electrodynamic model of the atmosphere–ionosphere coupling based on electromotive force that works between the Earth's surface and the near-ground atmospheric layers. Electromotive force is formed as a result of convective and turbulent transport and gravitational sedimentation of charged aerosols. The theoretical modeling of the atmosphere–ionosphere interaction at the preparatory phases of earthquakes can be applied to pulse release of the charged particles originating from armaments with DU used on the battlefields at the Earth's surface into the atmosphere. The facts related to electrical coupling of lithosphere-atmosphere-ionosphere may be considered in the functional and casual unity with the Schumann frequencies.

DU and the Schumann Resonance Hypothesis

A phenomenon, known as the Schumann resonance, was detected by satellite in space, well beyond the upper boundary of the resonant cavity which is formed by the Earth's surface and the lower edge of the ionosphere (Simões, Pfaff and Freudenreich, 2011). Random lightning strokes with spatial probability distribution peaking over the continents, particularly in the low latitude regions, induce development of standing waves whose wavelength is related to the radius of the cavity.

Many of these effects may be induced by a man-made ionospheric disturbance (Hainsworth, 1983). Numerous military actions, in which DU ammunition was used, occurred in low to middle latitude regions (Figure 1). The contamination of the environment, caused by man in the event of a nuclear war, assumes a release of the considerable amounts of different forms of uranium. Up to 70% of the DU penetrators are converted to aerosols. Lighter particles are globally dispersed (IPPNW, 2014). The observations of the authors mentioned above and our publications since 2013 (Zunic, 2013, Zunic, 2013-1, Zunic and Rakic, 2013, Zunic and Rakic, 2015) imply that repeated discharge of large amounts of the uranium oxide fumes from the battlefields in the Persian Gulf, the Balkans, Afghanistan, and other places, contributes to sudden artificial imprints of charged particles, resulting in the induced light pulses in the atmosphere. After colliding with high energy protons and nuclear fragments from cosmic rays, a "runaway process" can start inducing lightning initiation.

According to Fox (2011), much of the energy from the waves is trapped between the ground and the ionosphere layer in the Earth's atmosphere. These waves remain trapped inside an atmospheric ceiling created by the lower edge of the ionosfere – a part of the atmosphere filled with charged particles, which begins about 60 miles up into the sky. The waves created by lightning oscillate with regions of greater energy and lesser energy. Variations in the resonances correspond to changes in the seasons, solar activity, activity in the Earth's magnetic environment, in water aerosols in the atmosphere, and other Earth-bound phenomena. In field theory, any disturbance that drives fields from their harmonious rhythm spreads out to disturb neighboring fields. Our brain waves share and are attuned to certain frequencies of the Schumann's resonances, the Extremely Low Frequency (ELF) signals that pulsate between the Earth's crust and ionosphere (Miller, 2013).

We postulate that lightning flashes originating from the DU particles' radioactive decay in the atmosphere generate low frequency waves that circle around Earth, mingling with the waves that create a phenomenon, known as the Schumann resonance.

Studies on bioeffects on humans exposed to the electromagnetic pollution are important, especially in the context of Extremely Low Frequency Electromagnetic Fields (ELF-EMFs). Disturbances in human brain emitting theta and alpha frequencies in the same EMF region, or specific frequencies that trigger the production of different endorphins, beta-endorphins, catecholamines, enkephalins, dynorphins, proteins, and stem cells can undermine the basic integrative regulatory system in the human body, or more general, in the biosystem. The main health hazards of airborne particulate pollutants are cardio-respiratory disease and lung cancer (AGNIR, 2004). Electrophysiological considerations suggest that the central nervous system is potentially susceptible to induced electric field (WHO, 2007). The CNS in vivo is likely to be more sensitive to induced low frequency electric fields and currents than in vitro preparations (Saunders and Jefferys, 2002). EMF effects on the heart could theoretically result from indirect effects mediated via the autonomic nervous system and CNS (Sienkiewicz, 2003). Effects on the endocrine system could potentially also be mediated in this way (WHO, 2003).

The contamination of the environment, caused by nuclear wars, induces an unpredictable expression in the atmosphere. The lightning is the major source of electromagnetic radiation in the ELF range (Simões, Pfaff and Freudenreich, 2011). A study coauthored by Schaefer, Bourland and Nyenhuis (2000) described the existence of time-varying magnetic fields induced by electric fields that could interfere and change the biorhythm pattern. By analogy, we suppose that ELF-EMFs can be induced by artificial imprints of the unpredictable amounts of charged particles into the atmosphere during (and after) nuclear conflicts, and that these magnetic fields could interfere with living matter. Nuclear wars may be considered as very rare events, but electromagnetic changes induced by the imprinting of charged particles in the Earth's mantle have power to result in considerable physical phenomena (light, gradient magnetic field etc.). This approach opens an opportunity to observe one of key phenomena, the Petkau effect, in relation with the field of low-dose radiation, in a more comprehensive way.

Chapter 3

The Petkau Effect

In 1972, a researcher in Canada, Dr. Abram Petkau, found that when cells were irradiated slowly, a smaller total dose was needed to cause damage. This discovery is known as the "Petkau Effect". Dr. Abram Petkau discovered that at 26 rads per minute (fast dose rate), a total dose of 3,500 rads is required to destroy a cell membrane. However, at 0.001 rad per minute (slow dose rate), only 0.7 rads is necessary to destroy the cell membrane (according to Borthacur's web publication).

Although the Petkau effect was described in the literature as an *in vitro* phenomenon (Burlakova, 1999), the repeated bombing of relatively close areas in the Persian Gulf and the Balkans, with subsequent emission of ionizing radiation and a prolonged release of alpha particles, emitted during radioactive decay of DU, which originates from corroded DU armaments, provides an opportunity for the estimation of the *in vivo* Petkau phenomenon and its effects (Figure 5) (Zunic, 2013).

The low doses (air pollution easy transferable to the remote distances from the place of explosion) and the slow doses (depleted uranium ammunition remnants can be fully oxidized into corrosion products twenty-five to thirty-five years after the impact) have ensured a further, prolonged contribution to the maintenance of alpha particle radiation, leading to the consequent, disastrous Petkau effect in the biosphere (Zunic, 2013; Zunic and Rakic, 2013, 2015). Repeated, low-slow doses could induce remarkably harmful effects in living tissues, which reflects in the anticipated Petkau effect, with superimposing increasing trend over time, opposite to radiation which caused it.

There is a vast body of data related to the quantities of DU that was used in military campaigns since 1990 (WISE, 2011). The contours in Figure 5 reflect the literature data on the quantities of DU used in military campaigns in the Balkans and the Persian Gulf. Along with the expected reduction of radiation in the course of time, the Petkau effect is intensified. The most intriguing two questions regarding the Petkau effect are: 1) Is the Petkau effect a particle-induced phenomenon? 2) Is the Petkau effect an *in vitro* or may be an *in vivo* manifestation, too?

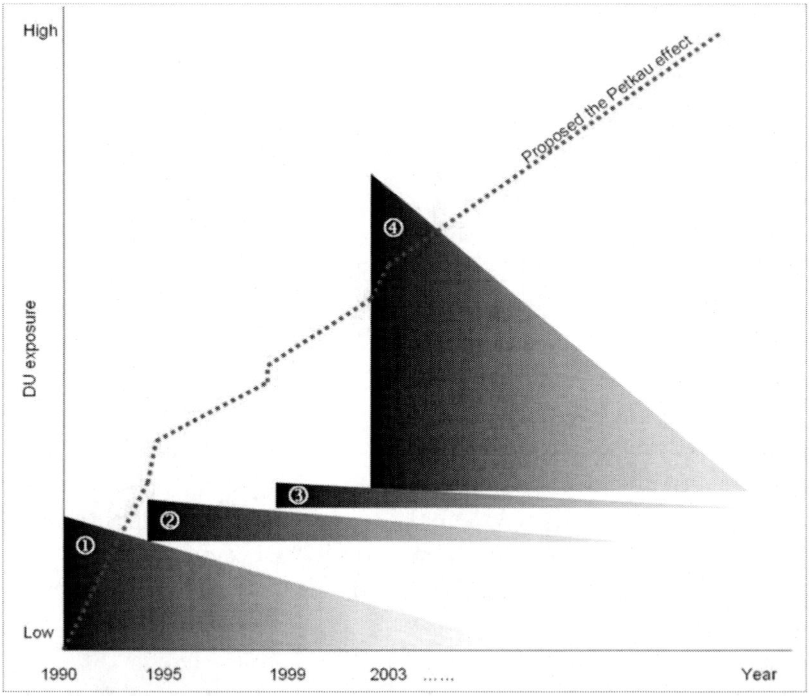

Figure 5. Presentation of the proposed *in vivo* Petkau effect in relation to repeated exposures to depleted uranium (DU). The dark gray to pale grey contours represent proposed exposures to alpha, beta and gamma radiation that originated from depleted uranium missiles in military campaigns: ① First Gulf War (1990-1991), ② bombing of Bosnia and Herzegovina (1994-1995), ③ bombing of Serbia (1999) ④ Second Gulf War (2003-2011); different size of the shaded areas symbolizes the difference in quantities of depleted uranium armaments used during military campaigns. The black dashed line symbolizes the proposed *in vivo* Petkau effect. Adapted from primary data of Zunic, (2013).

A long term exposure to low-dose radiation was found to be dramatically more harmful than a short term exposure to high level radiation. Reactive oxygen species (ROS), which are products of radiolysis of water, induce indirect low-dose radiation effects. These indirect effects are dominant cause of biological effects (Djurovic, Spasic-Jokic and Djurovic, 2008). The authors found that occupational exposure to low-dose ionizing radiation compromised mitochondrial function. One of the mechanisms is the damage of organelles by peroxidation of cell membranes, which increases when the dose rate is decreased (Djurović, Selakovic and Spasic-Jokic, 2004).

The biological effects of DU particles deposited in the body are perceived as the "ideal killer". Besides a biological predisposition, the physicochemical characteristics of radioactive particles are substantial for an understanding of "hot" and "cold" sequels of depleted uranium. With the possibility of observing a wave phenomenon in the atmosphere, with high impact on the biosphere, we can conclude that the Petkau effect could partially be a wave phenomenon, generated mainly at the membrane level disorganization.

Focusing our attention "back" from a macro plan on the micro world or *in vitro* conditions, we believe that low-dose radiation is more harmful than high-dose radiation, because the latter results in the irreversible harmful effects targeting cells and subcellular structures. The low-dose radiation, originating from the charged particles or photonic radiation can interfere with the biofrequencies of cell structures, primarily the cell membrane or the endomembrane system, which results in higher cytotoxicity.

In higher mammals, including humans, the Petkau effect is considered as a long term, but a timely-framed modification of the biosystems' integrative adaptive response, which includes the whole metabolome and the changed signaling pattern of regulatory mechanisms in "horizontal", as well as in its "vertical" organization.

We have proposed that the Petkau effect is induced by ROS as well as by wave effects, which most likely work in synergy at the cell membrane and/or DNA. Presentation of the interdependence of the proposed environmental effects influencing the biosphere and the postulated Petkau effect is shown in Figure 6.

To conclude, the use of large amounts of DU in the military actions in the Persian Gulf and the Balkans, almost every four years since 1990, resulted in the contamination of all natural resources, including ground and the atmosphere. The propagation of ionization events in the environment contributes to further interaction of charged particles and generation of magnetic fields which interfere with living systems. Owing to the unity of the lithosphere, atmosphere, ionosphere and biosphere, the interference of non-ionizing radiation fields, including the Schumann resonance, with biofrequencies occurs.

We observe the Schumann resonance and the Petkau effect as possibly synergistic, or at least, simultaneous events, which act at the biosystem integrative level.

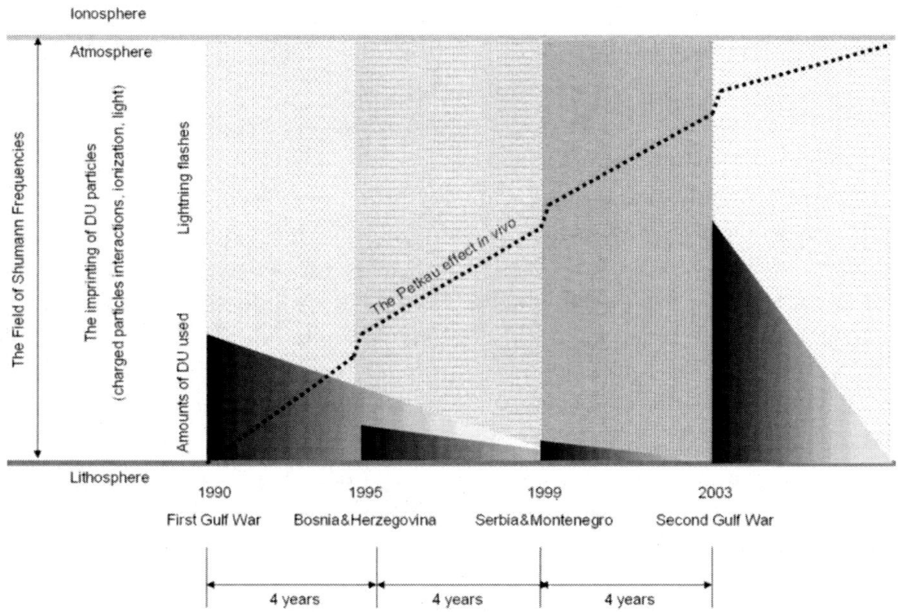

Figure 6. The impact of the military use of DU on the lithosphere-atmosphere-ionosphere and biosphere coupling. Light effects in the atmosphere are in relation to the amounts of depleted uranium used for military purposes.

Chapter 4

RADIATION AND BIOSYSTEM

The biosphere is composed of biosystems. Biosystems are made up of biotic and abiotic components. The biosystem's hierarchical organization means interaction with the physical environment (energy and matter) at each level, which produces characteristic functional systems (Odum and Barrett, 1971). After the Hiroshima and Nagasaki experience, the best scientific evidence of radiation effects on humans came unfortunately from epidemiologic studies of atomic bomb survivors (Pollycove, 1998).

What is the mechanism of the repeated low-slow radiation deleterious effects? Is it possible to apply an ancient Chinese proverb referring to the perception of the power of the weak? For example, soldiers must break step when crossing a bridge due to the risk of creating large motions at resonant frequencies. These rhythmic oscillations would reach a maximum peak when the bridge can no longer sustain its own strength and hence collapses.

In analogy with the above mentioned, cells are individual biosystems, which communicate and make up a higher level biosystem – a tissue with peculiar functional architecture (Odum and Barrett, 1971).

We have already discussed the hypothesis on the lithosphere-atmosphere-ionosphere and biosphere coupling and the hypothesis on the wave nature of the Petkau effect.

The first argument in favor of the hypothesis on the wave nature of the Petkau effect is based on a historical experiment performed by Petkau with the Na-22 radiation impact on cell membranes (Petkau, 1972). The Na-22 decay spectrum shows a gamma-ray at 1274.5 keV, an annihilation peak at 511 keV (from the β^+) and x-rays from the electron capture (data available online at http://ns.ph.liv.ac.uk/~ajb/radiometrics/glossary/sodium22.html). Gamma rays belong to the electromagnetic spectrum. They are highly energetic waves (energy limit > 2×10^{-14} J, with wavelength: < 1×10^{-11} m, and frequency: > 3×10^{19} Hz) (NASA 2013, NASA, 2013-1).

The second argument rests on the interaction of electromagnetic radiation with the information structures of living cells - nucleic acids, proteins and membranes. The effects of cytotoxicity are inevitably associated with the changes in metabolic and signaling properties of the cell.

The existence of the interfering radiation in the environment could disturb the information system of the organism, which depends on the initial state of an organism (Reshetnyak et al., 1996). The authors postulate that the external irradiation of cells with coherent waves may lead to the formation of substructures on the membranes of healthy cells and induce the corresponding generation of coherent waves. Sharp deformations of the membrane correspond to disturbance of the cell functioning and relate to the generation of coherent waves at resonance frequencies. The membrane shape changes, which lead to changes in the cell functioning, occur relatively fast.

The above mentioned argument may be the third one in understanding the (short) time dependent membrane deformability in case of environmental radiation that interferes with the cell.

The presence of *molecular antennas* in the cell creates biosystem's response to incoming waves. Several biomolecules, including enzymes (with active sites surrounded by biopolymer chains folded into globules), or chlorophyll, hemoglobin, myoglobin, have similar architecture: these molecules may have an ion which is situated in their geometrical centers (a magnesium ion in chlorophyll and an iron ion in hemoglobin). This ion is arranged at the center of symmetry of a tetrapyrrole ring with pseudoplane structure. The reaction ability of biomacromolecules substantially depends on the level of excitation of central subunits (Reshetnyak et al., 1996). An associating center (a metal ion) actively interacts via biochemical bonds with peripheral acceptors that receive encoded energy. Metaloenzymes have similar properties and they are included in important biochemical processes. At the same time, chlorophyll (in plants), hemoglobin (in erythrocytes) and myoglobin (in muscle) are oxygen carriers. The third class of biomacromolecules corresponding to the antenna model are macrocycle compounds (Reshetnyak et al., 1996). Porphyrins are a group of heterocyclic macrocycle organic compounds, composed of four modified pyrrole subunits (available online at http://www.shutterstock.com/pic-355192466/stock-vector-porphyrins-are-a-group-of-heterocyclic-macrocycle-organic-compounds-composed-of-four-modified.html).

There are many examples of vitally important proteins that incorporate one or more metal ions within their structure: enzymes, proton transport proteins, electron transfer/transport proteins, storage proteins, metallochaperones, iron transport proteins, oxygen storage/transport proteins, calcium binding proteins, monooxygenases… The metal coordination environment is often generated from residues inherent to the protein, small exogenous molecules (e.g., aqua ligands) and/or macrocyclic porphyrin units found, for example, in hemoglobin, myoglobin, cytochrome C, cytochrome C oxidase, and vitamin B_{12} (Joshi, Graham and Spiccia, 2015).

Figure 7 shows that the incoming radiation may excite electrons in cell molecular structures.

An energy diagram shows how an electron in the biomolecule can be elevated to a higher energy level (according to Koning, 1994). Its de-excitation to the stable excitation state is responsible for heat production. Consequently, heat shock proteins (HSPs) can be expressed in the local tissue microenvironment.

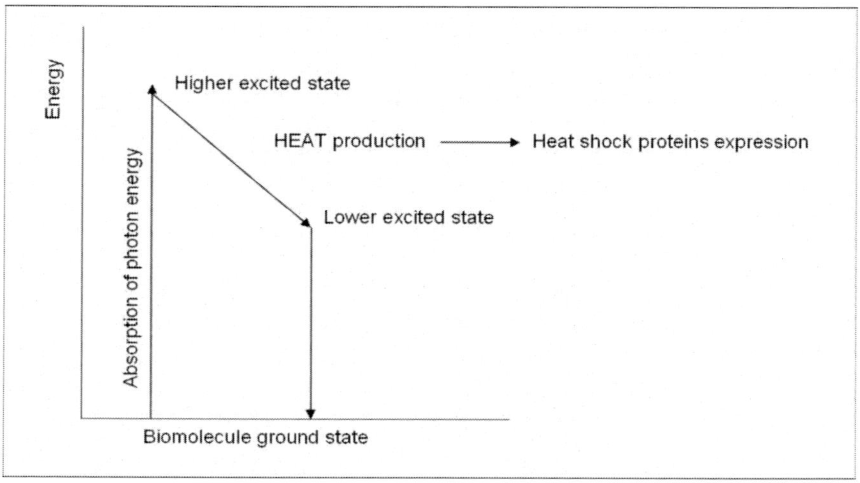

Figure 7. The absorption of photons by cell molecular structures.

DNA, as well as other biopolymers, is a potential energy absorbing molecule within the cell. We have mentioned above that HSPs are expressed in an anti-stress response under the influence of increased environmental temperature, which contributes to the causal connection between the atmosphere and biosphere (according to Figure 4 and Zunic and Rakic, 2013). HSPs are involved in steroid hormone signaling. Steroid hormone receptors may contain a HSP sequence. Steroid hormone receptors are found in the cytoplasm and in the nucleus. Upon entering the cell, HSP dissociates from the complex with steroid hormone receptor. Receptor-ligand transport into the nucleus is facilitated by a nuclear localization signaling sequence. This sequence is exposed only when HSP dissociates (Leeper-Woodford and Adkison, 2016). Steroid and thyroid hormone receptors are transcription factors which modulate gene expression in target cells (Bowen, 1998).

Electromagnetic fields activate the cellular stress response, a protective mechanism that induces the expression of stress response genes. The HSP70 promoter shows two different DNA sequences that have been identified as activated by EMF (non-thermal) and by thermal stimuli. The studies with model biochemical systems suggest that electromagnetic field could interact directly with electrons in DNA. Induction of increased levels of the major stress protein, HSP70, by EMF is rapid, within 5 min, and it occurs at extremely low levels of energy input (14 orders of magnitude lower than with a thermal stimulus) (Blank and Goodman, 2009). The cell heat-shock effects are associated with reversible phase transitions of DNA liquid crystals in chromosomes during heating and cooling in the sublethal temperature ranges from 40-43°C (Reshetnyak et al., 1996).

What happens if the HSP expression is high enough to disturb the signaling mechanism of steroid hormones? Is it possible to achieve a “futile” spin in which a stress response molecule may be interpolated in the stress response signaling mechanisms mediated by thyroid and steroid hormones? Provoked transcription regulation may result in all levels of endocrine deregulation. If stress stops, then the revitalization of entire signaling can be achieved. If persists, then, depending on the cell type or radiation susceptibility, a pathogenic process may develop.

This speculation confers “vertical” (hierarchical) meaning to the Petkau effect induced by DU, which actually acts as an endocrine disruptor.

Electromagnetic fields can impair the flow of protons through ATP-synthase Fo. Russian physicists (Semikhina and Kiselev, 1981; Semikhina et al., 1988) reported that very low levels of ELF MFs (25 nT) could alter the structure of water, and that the effects of the altered water structure would be particularly important under high concentrations of protons and water molecules. The consequence is diminished ATP production, due to altered function of ATP synthase in inner mitochondrial membrane. Decreased ATP level triggers a cascade of events, starting from the activation of AMP-activated protein kinase (AMPK), the enzyme which is extremely sensitive to changes in ATP level.

At the same time, ATP molecule is a key adapter in the chain of transformation of energies and the ATP cycle. ATP deposes metabolic energy (in form of highly-energetic phosphoanhydride bonds) and influences cell energy charge which must be tightly regulated to provide optimal cell homeostasis (Zunic, 2016).

If the pathogenic effects of radiation are repeated or long-lasting, then, metabolic regulation changes, and confers to mainly "horizontal" meaning of the Petkau effect induced by DU. Depleted uranium is, in fact, a metabolic disruptor.

The emission of radiation, in the course of several decades, due to corrosion of the scattered remnants of DU armaments, and which has been intensified by the repeated bombing of the regions within the range of the transfer of radioactive particles in the air, strikes a broad territory and numerous populations, and unavoidably leads to the *in vivo* Petkau effect. Except the war, peacetime nuclear disasters in various parts of the world, such as Fukushima, Chernobyl and others, contribute to this effect, too. In this way, the Petkau effect is a challenge for science to declare the future health strategy with the main goal focused on minimizing the early, as well as delayed *in vivo* effects of depleted uranium (Zunic, 2013).

The primary objective of this study is to contribute to a better understanding of the interaction of depleted uranium, as a source of low dose radiation, with the living world and man in contaminated environment. We viewed living systems in the unity of microsystems, as the cell is, through the complex hierarchical integrative regulation.

Man is the main subject of our study. Understanding of the basic principles of cell biology and radiation interaction with living matter was supported by authentic medical data. These data were obtained from pediatric patients who were examined for clinical symptoms and radiological changes consistent with persistent pulmonary infiltrates in the first year after the bombing of Serbia and close geographic territories, and from pediatric and adult patients who originated from the territory exposed to DU air contamination since 1992. The fact that depleted uranium is the source of the long-lived isotopes, some of which have extremely long half-lives, and that uranium is an emitter of alpha, beta and gamma radiation, encourages our need to predict the further course of events both in nature and the living world. As our estimates become more precise and our knowledge increases, it also increases our chances of repairing the consequences of depleted uranium contamination. At the same time, we raise our awareness high enough to be able to provide highly professional standards in the management of the sources of ionizing radiation.

Routes of Depleted Uranium Entering the Body

As in the case of other metal-oxide nanoparticles, with higher temperature of explosion, the DU deliberated particle size is lower. The dimensions of DU particles are inversely proportional to their penetrability. There is an evidence of exposure to dispersion of a new type of uranium, the ceramic submicron oxide particles, especially in European countries closer to Iraq, than those in remote parts of Europe (Busby and Morgan 2006). Because the ceramic DU dust particles are not soluble, they remain in the body much longer than other soluble forms of uranium. The "Trojan Horse effect", described by Park et al. (2010), and "lysosome-enhanced Trojan horse effect" demonstrate the importance of the fine insoluble particles that, due to high penetrability, can cause harmful effects in the cell, facilitating entering of other toxic components, or interacting with cellular structures (Ortega et al, 2014; Sabella et al, 2014).

Even though uranium is a powerful genotoxic stressor, the health effects caused by DU radiation may not appear for years. The pathogenesis of nanoparticle-induced lesions, in the case of DU is more complex, due to the superimposing biological effects of α, β and γ radiation. DU is primarily an alpha emitter and inducer of a mixed radio-chemical exposure. Alpha particles have been shown to cause a variety of effects on cells, including cell death, gene mutation, chromosome aberrations, and malignant transformation leading to cancer. Owing to their relatively large size, alpha particles collide readily with matter and lose their energy quickly, along a short track, inducing high ionization. Beta particles and gamma rays also damage cells. Beta particles have the medium penetrating power and the medium ionizing power. Gamma rays have great penetrating power and can pass through the human body (Military Toxics Project Information Sheet, 2003).

The effects of DU on human health depend on the types and magnitudes of exposure, as well as on characteristics such as particle size, chemical form, and solubility (Burger, 2012). DU nanoparticles can penetrate multiple tissue and cell structures. According to the results of Gatti and Montanari (2004), the particles found in the places of its primary deposition after entering the body, lungs and the stomach, were larger than those detected in other organs. After embedding into the tissue, DU micro or nanoparticles exert toxic effect of heavy metal as well as radiogenotoxic effect of mixed α, β, γ radiation. Once internalized by the cell, metal-oxide particles exert cytotoxic effect that is inversely proportional to the particle size (Gatti and Montanari, 2004).

Exposure to depleted uranium can occur by inhalation of DU dust, ingestion of DU directly, or in contaminated food, soil, and water, embedding of DU fragments in the body, contamination of open wounds with DU dust, and absorption through contact with the skin (Military Toxics Project Information Sheet, 2003).

Nanoparticles can cause irreversible damage to cells by oxidative stress or/and organelle injury, preceding tissue inflammation (Limbach et al, 2007) and the altered cell death mechanisms (Buzea, Blandino and Robbie, 2007).

Reactive oxygen species may also be generated as a result of nanoparticle-cell interactions (Xia et al., 2006). When the generation of ROS exceeds the cell's antioxidant capacity, a state of oxidative stress is induced in the cell. Experimental studies showed that local persistence of low toxic particles in the lung results in chronic inflammation that may be a cause for non-genotoxic induction of lung tumors (NIOSH, 2011). In experimental conditions, repeated exposure to DU induces a causal cascade of related events, where DNA-strand breaks in the bronchoalveolar lavage (BAL) cells occur due to impaired tissue immunoreactivity, infection and ROS production (Miller et al., 2002).

Polluted food implies lymphogenic dissemination of nanoparticles (Gatti and Montanari, 2004). Over time, DU particles will be swallowed or absorbed by the body, but some may remain in the lungs for years. DU absorbed into the blood will be excreted from the body, primarily in urine, while DU taken into the gut will be excreted in feces. Up to 75% of DU absorbed into the blood may be excreted during the first week, followed by slow excretion for up to a year. DU will be deposited in bones and organs, especially the kidneys (WHO, 2001). DU will remain in the kidneys for at least three months and in bones for at least twenty-five years (Military Toxics Project Information Sheet, 2003).

The toxic properties of DU primarily affect the kidneys. DU may disrupt kidney function, or in high doses even cause renal failure, as data suggest, based on the research involving soluble uranium (which is mostly a renal toxicant), rather than on insoluble uranium such as DU dust (which is a lung chemical toxicant and systemic radiological hazard) (Gatti and Montanari, 2004).

The inhalation of DU particles is the most common path of internal contamination (Gatti and Montanari, 2004). The inhaled DU dust will settle in the nose, mouth, airways, lungs, and gut. Experimental and human studies have showed that the macrophage clearance mechanisms in the lungs remove inhaled nanoparticles less efficiently than larger particles. The tissue embedded particles of DU may induce or aggravate inflammatory and allergic responses by directly influencing lung immunocytes (Inoue and Takano, 2011).

The nature of alpha-particle induced radiobiology effect is dependent on the uniformity of tissue distribution of α-emitting radionuclides (Muggenburg et al., 2008). Alpha particles deposit the dominant doses to the lung, but also comparable doses to the thoracic lymph nodes, and smaller doses to the extra-thoracic airways and lymph nodes, bone surfaces, kidney, liver, and red bone marrow. The interaction of α-particles and thoracic lymph nodes with bystander effect (BSE), as a main pathogenic trigger, changes the T-cell immunobiological properties, and T-cell may be directed to the central nervous system, as well as to other tissues. The lung can serve as a location of interactions between alpha particles and the lung immunocytes, where autoreactive T cells become reactivated and gain the competence to enter the CNS. The lung could therefore contribute to the activation of potentially autoaggressive T-cells and their transition to a migratory mode, as a prerequisite to entering their target tissues and inducing autoimmune diseases (Odoardi et al., 2012).

Depending on aerosol speciation, inhalation may lead to a protracted exposure of the lung and other organs (Bleise, Danesi and Burkart, 2003). Particles smaller than 10 μm can intrude deep into the lung. Particles smaller than 100 nm have a tendency to be deposed in alveolar space. As particles are smaller, their deposition rate is higher (Ulrich et al., 2011). Small and insoluble metal-oxide particles penetrate the alveoli into the circulation from which they are rapidly distributed into the entire organism. Tissue penetration from alveoli to the blood vessels is highly particle-size dependent. Particles the size of 100 nm, when inhaled, enter the blood flow within 60 seconds and can be found in internal organs in a matter of minutes (Gehr and Heyder, 2000).

The metal-oxides can cross the blood-brain barrier. These particles also cross the placenta into the fetus (Ulrich et al., 2011). Experimental exposure to DU led to impaired coordination and movement performance in rats with multisystem damage including the brain (Seideman et al., 2011).

The radiotoxic effects of depleted uranium with concomitant alpha particle radiation has been associated with unpredictable and everlasting biological effects.

Since 1991, the territory of the Republic of Serbia has been repeatedly exposed to radioactive DU particles. Figure 1 is based on the reported data concerning the detection of radiation in the remote regions of Europe, which coincided with the operations in the Persian Gulf, Bosnia and Herzegovina and Serbia. According to the authors' civilian access to data and theoretical interpretation of an observation given by Busby, Morgan (2006) and Moret, (2006), it was possible to estimate an approximate 2,400 mile radius around the Persian Gulf, as well as around Bosnia and Herzegovina and Serbia, with the putative expansion of air pollution containing DU particles (Figure 1), but without taking into account any geographic or meteorological peculiarities of the potentially exposed area (Zunic, 2013).

It is obvious that internal contamination with DU particles leads to unforeseeable systemic effects with complex pathogenesis. Cancer lesions arise from tissues with impaired metabolic and immunological competence. An achievable increase of subsequent primary cancers (Levine et al., 2005; Bradford et al., 2010) supports the thesis that aggressive T-cells may be scattered away from lung lymph nodes (Odoardi et al., 2012).

The literature data show that the effects of α-emitting particles inhaled in experimental conditions have delayed oncogenic potential. Alpha radiation induced lymphopenia, atrophy and fibrosis of the thoracic lymph nodes, radiation pulmonary fibrosis. These tissue lesions occurred before cancer lesions, in a prolonged course. Due to the bystander effect, the oncogenic transformation after irradiation of 10% of the cells in culture is the same as all the cells on the dish are exposed to the same number of α-particles (Sawant et al., 2001).

A vast amount of data has underlined the multisystem nature of the body response to the effects of DU. Gatti and Montanari (2004) investigated twenty cases of Italian soldiers and 8 cases of civilians living in Sarajevo at the time of the bombing of Bosnia and Herzegovina (1994-1995). Although in some tissue samples the presence of DU has not been proven, the possibility that the patients in the exposure zone were internally contaminated cannot be ruled out, as well as the presence of DU particles in their bodies.

Chapter 5

RADIOBIOLOGY OF DEPLETED URANIUM

The environmental impact of DU, as well as numerous radiobiological effects caused by DU exposure, has become evident during the last 25 years, in parallel with the increasing frequency of the local conflicts and military use of DU.

Our interest in radiobiology of DU has been aroused owing to the ever more present use of biologically significant radionuclides including the man-made, as well as naturally occurring radioisotopes (Brack, 2011). According to Tables 2 and 3, and a publication by Brack (2011), some radionuclides are characterized by high toxicity from the U-238 decay series: Th-230, Ra-226 and Pa-231 and Ac-227 from the U-235 decay series. Radionuclides of medium toxicity are U-238, U-234 and Bi-210 from the U-238 decay series and U-235, Th-231, Th-227 and Ra-223 from the U-235 decay series.

DU exhibits a *heavy-metal & radiation* synergic impact on the biosystem. The use of DU, or release of toxic and ionizing substances during peacetime nuclear disasters, induces hormetic effects on the environment, as well as on living organisms (Zunic and Rakic, 2013). Both, high chemical toxicity of DU, as well as its radiation, can damage genetic and non-genetic material in the cell (Munroe, 2004). The alpha track is highly destructive to DNA and other biomolecules (Tartier et al., 2007). DNA damage was partly a consequence of the inflammatory processes and ROS production. Repeated exposure to insoluble DU particles could induce a potentiation effect (Monleau et al., 2006).

We believe that getting an insight into a few key concepts can help us better understand the impact of man-made radionuclides on the natural resources' fragile balance. Regardless of whether they get into the environment after nuclear accidents, or after military actions, the man-made radioisotopes contaminate the environment. The hypothesis of limited contamination after the use of nuclear weapons has been undermined by loads of evidence that we mentioned in our recent publications (Zunic, 2013-1; Zunic and Rakic, 2015). Busby, Hamdan and Ariabi (2010) explained that exposures to uranium oxide nanoparticles from weapons do not pose the same kind of hazard as uranium exposures in people living in high background uranium areas.

Close to the battlefields where the blasts of projectiles with DU lead to direct contamination of the external environment and people, causing exposure to higher doses of radiation, deliberated micro or nano particles of DU, in form of air pollution, are easily transferable to the remote distances from the place of explosion (Burger, 2012). DU induces prolonged, low-slow dose radiation in wide population, at global level (Burger, 2012; Zunic, 2013; Zunic and Rakic, 2013).

This fact allows the authors to highlight the “inversions”, that is to say, the demonstration of power by those less powerful: 1) Low-slow doses induce more harmful effects on the cell than higher doses rates (the Petkau effect); 2) Unirradiated cells exhibit irradiated effects as a result of signals received from nearby irradiated cells, which is the definition of the “bystander effect” (BSE).

Natural background radiation varies from ~0.01 mSv/day to 5-fold higher values. Acute exposures to high-dose radiation of >150 mSv, causes immediate, measurable and often serious effects on humans. Low-dose radiation is between background and high-dose radiation. Exposure to low-dose radiation has no immediately noticeable effects on humans (Bonner, 2003).

There has been increased interest in biological effects of low dose radiation after Chernobyl, military use of nuclear weapons, including DU, and other nuclear catastrophic accidents (Baverstock, 2000). The military use of DU for decades put the problem of low-dose radiation exposure in the spotlight. That is why one of key tasks of radiobiology is to provide better insight into the mechanisms of biological effects of DU.

The uncertainty of epidemiological studies about the health effects of low-dose radiation arises from the fact that the biological effects of low-dose radiation do not relate obligatory to DNA damage (Morgan, 2003; Morgan 2003-1; Il'yasova et al., 2014). DNA in active genes is repaired with higher efficiency and faster than in silent genes (Oleinick, Chiu and Friedman, 1984; Bohr, 1987). In case of high dose radiation exposure, DNA damage is a main contributing factor of carcinogenicity (UNSECAR 2006).

MODELS OF TISSUE RESPONSE TO LOW DOSE RADIATION

According to a publication authored by Il'yasova and coworkers (2014), currently, a linear no-threshold model (LNTH) is used to estimate health risks associated with exposure to low-dose radiation. Low radiation doses (<0.2 Gy) with a significant uncertainty relate to risk of malignant and nonmalignant diseases (Preston et al., 2013). High-dose ionizing radiation (>1 Gy) has a linear relationship to cancer risk in humans (IARC, 2012).

Confusing and contradictory conclusions regarding dose-response relationship depend on what is emphasized, epidemiological data, or biological mechanisms (Ulsh, 2012). There are many models of low-dose radiation risk assessment: 1) The linear no-threshold model is a model used in radiation protection, assuming that any level of radiation is harmful. There is a linear risk increase in parallel to dose increase. The conventional approach extrapolates the linear trend observed at high-dose radiation to low-dose exposures, resulting in the linear no-threshold hypothesis (Il'yasova et al., 2014). This model also implies the bystander effect, which on the other side contributes to uncertainty of this model (Elgazar and Kazem, 2006); 2) The threshold model assumes the linear increase of risk related to dose, but only after a certain threshold level is exceeded. The threshold-model postulates that low-dose radiation is harmless below a certain level (Bonner, 2003; Elgazar and Kazem, 2006); 3) The hormetic model includes a bimodal effect of radiation: below a certain threshold it provokes radioprotective mechanisms, but above the threshold, radiation induces damaging effect (Elgazar and Kazem, 2006).

There is a possibility of combining the models of tissue response to low dose radiation. This paper will introduce the hormesis-threshold model of low dose-response, based on the neural network method, which is discussed bellow according to a publication by Zunic et al., (2007) and Zunic, (2015).

NON-TARGETED EFFECTS OF IONIZING RADIATION

The electrical properties of biological tissues are determined by the electrical interactions of polar molecules and ions (Formica and Silvestri, 2004). Ionizing radiation can cause discrete increases in the energy of a molecular or atomic absorber, causing irreversible alterations in atomic configurations resulting in ionization or disruption of covalent bonds (Krasin and Wagner, 1988).

Non-targeted effects, which are not a direct consequence of the initial lesions produced in cellular DNA, include: bystander responses, genomic instability, gene induction, adaptive responses, low dose hypersensitivity (Prise et al., 2002), clastogenic factors produced in plasma from irradiated individuals that can cause chromosomal damage when cultured with nonirradiated cells, and transgenerational effects of parental irradiation that can be manifested in the progeny (Morgan, 2003). Non-targeted effects dominate at low doses and saturate with increasing dose (Prise, 2002).

Non-targeted effects of radiation are time-evolving and can lead to delayed health effects, including cancerogenesis. This study will illustrate this challenge in the next chapter.

Radiobiology is fundamental to the estimation of environmental radiation risk, but cannot explain the phenomena of radiation induced genomic instability and the bystander effect (Baverstock, 2000).

According to the LNTH, the bystander effect hypothesis predicts a risk greater than expected because irradiated cells can transmit radiation effects to un-irradiated neighbouring cells (Il'yasova et al., 2014). Bystander effects have been observed *in vitro* and *in vivo* and for various radiation qualities (Morgan and Sowa, 2007).

Alpha particles emitted from DU particle that is trapped in the tissue cause ionization, whose main effect is the bystander effect in the living tissues. The radiation-induced bystander effect was first reported by Nagasawa and Little (1992). The bystander effect in radiation biology is explained as a response of cells that have not been directly exposed to radiation, but were in contact with, or received signals from irradiated cells (Acheva et al., 2014). BSE is predominant at low tissue doses; at higher doses, cell death is more probable than the stimulation of radioadaptive tissue response (Iyer and Lehnert, 2002).

The mechanisms of non-targeted effects of ionizing radiation include molecular signaling from irradiated cells, which propagates and induces the changed signaling pattern in non-irradiated neighbouring cells. One of the mechanisms includes a transfer of molecules through intercellular communication and the medium transfer system, which is leading to BSE. Due to the alpha-particle-induced bystander effect, the unpredicted cells close to irradiated cell populations can exhibit genetic alterations. BSE is partly a result of signaling molecules that rapidly propagate from irradiated cells and their concentration decreases (exponentially in the case of planar symmetry) as distance increases (Monleau et al, 2006). The mechanisms underlying non-targeted responses appear to involve production of reactive oxygen species and direct cell-to-cell signalling via gap junctional intercellular communication, although significant differences exist in different cell types. ROS can act as signal molecules in important cellular processes, including apoptosis (Lehnert and Iyer, 2002). Lorimore, Coates and Wright (2003) assume that BSE is mediated by gap junctions, cytokines and oxygen species. For signal propagation from irradiated to non-irradiated cells by gap junctions, the attachment of cells is necessary. In both cases (propagation through gap junctions, or diffusible molecules, including cytokines released from irradiated cells), the non-targeted effects of ionizing radiation are associated with free radical-mediated processes (Wright, 2004).

A lesion within supersensitive chromatin regions can trigger a cascade of metabolic processes that mark damaged cells to undergo apoptosis (Hall and Hei, 2003). BSE is important for elimination of highly damaged cells from the proximity to targeted ones (Belyakov et al., 2002; Prise et al., 2002). Cell death, including apoptosis, has been described as the principal response to low linear energy transfer (LET) (x- or gamma-ray) radiation exposure (Mothersill and Seymour, 2006).

A seminal paper authored by Hall and Hei (2003) is, as reported by the authors, perhaps "the most direct and most dramatic demonstration of the bystander effect that involves the observation of micronuclei (MN) in irradiated human fibroblasts". They showed experimentally that cells irradiated with 10 α-particles (in 48h incubation time), directed at the centroid of the nucleus, induced micronuclei and chromosome bridges to no-hit cells in fibroblast mixed culture. A study by Sawant and coworkers (2001) supported the need for gap-junction communication as a mediator of the bystander effects. Multiple mechanisms are involved in the bystander process depending on the cell types used, biological end points measured and types of radiation used (Hall and Hei, 2003).

Randers-Pehrson and coworkers (2001) reported that BSE was observed following cytoplasmic irradiation. BSE is detectable in cells that were not traversed by radiation, but were in the same environment as the irradiated cell (Kovalchuk and Baulch, 2008).

Experimental data demonstrate that effects of the parental radiation exposure are transmitted through the germline to the progeny of the irradiated parent (Morgan, 2003-1; 2003-2; 2003-3).

Environmental exposures during early development have a role in susceptibility to disease in later life. Epigenetic modifications provide a possible link between the environment and alterations in gene expression that might lead to disease phenotypes (Jirtle and Skinner, 2007). The authors reported that some of these environmental effects could be passed on through subsequent generations. In DU-exposed cells, Miller et al. (2002), showed that DU caused genomic instability through the formation of micronuclei lasting for at least 30 generations in the progeny of the cells first subjected to the low-level ionizing radiation and chemical toxicity. Genomic instability with its ability to allow genetic mutations and chromosomal aberrations to be passed on to future generations of cells may lead to cancer as well (Seideman et al, 2011).

The term "bystander" indicates a generally "horizontal", although blurred demarcation of biological effects of α-particles. Does the α-particle-induced bystander effect reflect hot or cold lesion, or a mix of both? Evolutionally conserved stress-defense mechanisms could be mobilized in a "vertical" manner, in addition to the bystander, "horizontal", mainly paracrine regulation. All levels of neuroendocrine feedback mechanisms may be altered, including heat shock proteins (Zunic and Rakic, 2015).

DNA damage and membrane permeability changes lead to loss of barrier function (Inan, 2005). Protein molecules and DNA need highly organized water surroundings of intracellular and intercellular medium to function properly. The resulting consequences of absorption of radiation by water molecules in biostructures may cause structural heating of water, phase transitions in biological membranes, etc. Large gradients of the electric field and related thermal gradients may arise in such a layer, even at low intensities of radiation. This may stimulate mixing of extracellular liquid, a change in transport of ions in water and other substances through cellular membranes, and give rise to thermal and acoustic waves, with no resonance character (Reshetnyak et al., 1996).

The alpha-particle-induced bystander effect was described as a result of the dynamic interaction between cells, assuming that the presence of DU particles, which are internally deposed in the body, induces radiogenotoxic changes to the surrounding cells (hot lesions) and inevitably influences the creation of systemic response and communication between the target tissue and distant tissues through the bloodstream or other body fluids (cold lesion aspect) (Zunic, 2013-1). Based on the litosphere-atmosphere-ionosphere-biosphere coupling hypothesis (Zunic and Rakic, 2013), in case of internal and/or external DU contamination, the conceptual presentation of interdependence between the Petkau effect and BSE is presented in Figure 8.

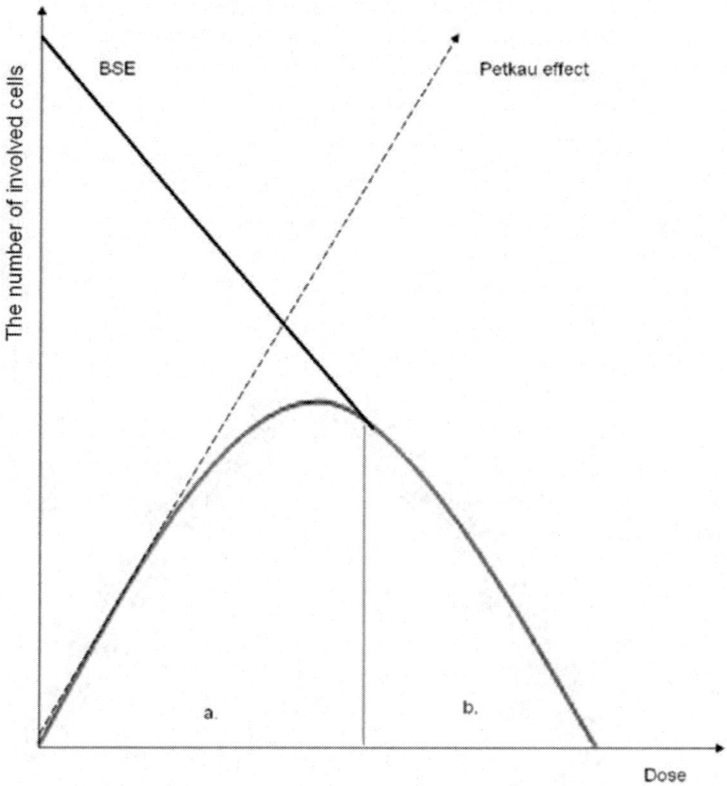

Figure 8. The conceptual presentation of interdependence between the Petkau effect and bystander effect (BSE). BSE is presented as a double-phase tissue response: a) phase related to radioadaptation (including hormetic tissue response); b) phase of exhausted radioadaptaton leading to irreversible changes, resulting in cell death.

Some non-targeted effects may act as protective mechanisms leading to the removal of potentially damaged cells from the population. BSE is presented in Figure 8 as a double-phase tissue response: a) phase related to radioadaptation (including hormetic tissue response); b) phase of exhausted radioadaptaton leading to irreversible changes, resulting in cell death. The curves presenting BSE and the Petkau effect exhibit the opposite trends. The Petkau effect-curve is more likely to exhibit exponential, rather than linear growth. The exponential growth of cytotoxic events may be achieved with repeated low-slow radiation dose exposure. The proposed concept asserts that the Petkau effect is a wave phenomenon, which makes its interdependence to BSE easier to understand.

The interaction of living matter with radiation may be caused by external, as well as internal sources. Time-dependent effects can be observed in case of ionizing or non-ionizing radiation, originating from external or internal sources (UNSECAR, 2000).

The biological effects of ionizing radiation depend on the dose. Higher doses may induce cell death. According to our estimates, the genetic changes of targeted cells are associated with higher doses, while non-DNA cell damage induces a wide spectrum of cell- molecular events, including an epigenetic modification. Otherwise, epigenetic modifications provide a spectrum of phenotypes providing better adaptation (Kovalchuk and Baulch, 2008).

Radiation-induced genomic instability is characterized by a number of delayed responses including chromosomal abnormalities, gene mutations and cell death (Coates, Lorimore and Wright, 2004). Epigenetic mechanisms are included in the evolving of the indirect radiation effects, including the radiation induced genome instability, BSE, and transgenerational effects (Kovalchuk and Baulch, 2008).

The central dogma of radiobiology explains that the damaging effects of ionizing radiation, including cell death, cytogenetic changes, apoptosis, mutagenesis, and carcinogenesis, are results of the direct ionization of cell structures, particularly DNA, or indirect damage via water radiolysis products (Widel, Przybyszewski and Rzeszowska-Wolny, 2009).

Ionizing radiation is a potent DNA damaging agent capable of producing, for example, cross linking, nucleotide base damage, and single and double strand breaks (Little, 2000). One of the starting paradigms of radiobiology is that direct damage to DNA is a prerequisite of heritable biological effects. Biologically effective linear energy transfer (LET) depends on coincidence between the DNA geometry (diameter of the DNA helix) and incoming ionizing events. The most damaging biological effects are in case of neutrons of a few hundred kiloelectron volts, as well as low-energy protons and α-particles (Hall and Hei, 2003).

Ionizing particles and rays collide with molecules in ~1% of the 100 trillion cells that make up the average human. These collisions generate clusters of free radicals that randomly damage cellular constituents including DNA (Bonner, 2003). About two-thirds of biological damage caused by LET radiation is due to indirect action (Elgazar and Kazem, 2006).

DNA damage induced by the direct effect of radiation is mediated through two mechanisms: 1) direct ionization of DNA; 2) "quasi-direct" effect which refers to the ionization of that portion of the solvent shell that is tightly bound to DNA and rapid transfer of holes and electrons created in the DNA solvation shell to DNA. This effect can be observed as an indirect effect and the initial one by its nature. It is impossible to differentiate this kind of DNA damage from direct DNA damage. The indirect-type damage is due to reactions with hydroxyl radicals and aqueous electrons (Swarts et al., 2007).

Double-stranded DNA (dsDNA) is a semiflexible biopolymer. Although dsDNA hybridization is stable at room temperature, local melting in the dsDNA may occur through thermal fluctuations. At high temperatures, DNA melting is facilitated by increasing the entropy of DNA (Kim et al., 2015).An insight into DNA bending under high stress could lead to a further understanding of many biological processes, including incoming radiation which triggers local ionization events.

Regardless of dose, whether the source of radiation is natural or man-made, radiation affects cells inducing direct and/or indirect effects. The ionizing radiation induced lesions of DNA include single strand breaks (SSB), base lesions, sugar damage, and apurinic/apyrimidinic sites (AP sites). Clustered DNA damage, a unique feature of ionizing radiation, is defined as two or more lesions within one to two helical turns of DNA, induced by a single radiation track. A double strand break (DSB) is a type of clustered DNA damage, in which single strand breaks are formed on opposite strands in close proximity (Shikazono et al., 2009).

The direct-type base damage is the trappable-radical single-track pathway. There is a balance between products derived from electron loss and electron gain. The ratio of base damage to deoxyribose damage is ~ 3:1 (Swarts et al., 2009). Oxidation of the two purines, guanine and adenine, are two of the three sites from which statistically significant yields of oxidation products were identified. The third site is deoxyribose. Guanine has the lowest oxidation potential and this base is expected to be the major site of hole capture (Swarts et al., 2007).

Gene expression is a very sensitive indicator of radiation exposure (Coleman et al., 2005, Yin et al., 2003). There is a difference in gene response to high and low-dose radiation (Amundson, 2003; Mezentsev and Amundson, 2011; Pouget et al., 2015).

Lowe and coworkers (2009) concluded that the cellular damage response mechanisms after low-dose radiation were qualitatively different from high-dose mechanisms. Wyrobek et al., (2011) investigated the low-dose dependency of the transcriptional response of human cells to characterize the shape and biological functions associated with the dose–response curve and to identify common and conserved functions of low dose expressed genes across cells and tissues. A set of about 80 genes that were associated with homeostasis mechanisms like membrane signaling, molecule transport, endomembranous system, including Golgi, and endoplasmic reticulum (ER), and signaling pathways showed consistent responses to graded doses of radiation spanning the range of 1–10 cGy. Transcription of MYC, FOS and TP53 was modulated at doses below 1 cGy in human lymphoblastoid cells. Jin et al. (2008) focused on the transcriptional responses induced by low and very low doses of ionizing radiation with time effect. Among 10,800 investigated genes, the genes that showed dose-dependent expression responses were involved in signal transduction, regulation of transcription, proteolysis, peptidolysis and metabolism.

All investigated genes were divided into groups of time-dependent and dose-dependent genes:

Dose-dependent genes were mainly associated with the signal transduction pathway, the regulation of transcription, proteolysis, peptidolysis and metabolism. When clustered, 22 genes showed early up-regulation to 0.01 Gy, but little response to other irradiation doses. The observed response was down-regulated at 48 h after exposure. These genes regulate carbohydrate metabolism, cell adhesion, cell proliferation, DNA-dependent regulation of transcription, G-protein coupled protein signaling pathway and other intracellular signaling cascades. Another cluster contained 16 genes showing down-regulation after 12 h of exposure to 1 Gy, but no response to other doses. These genes were functionally associated with phosphorylation, control of skeletal myogenesis and signal-dependent regulation of myogenesis by corepressor MITR.

Time-dependent genes were divided into two distinct groups: 1) the up-and-down group contains genes that are associated with cellular defense mechanisms such as apoptosis, cell adhesion, stress response and immune response and 2) the down-and-up group of genes, which relates to fundamental cellular processes including DNA replication, mitosis, RNA splicing, DNA repair and translation initiation. Genes showing both dose-and time-dependent responses exhibited a mixture of both features.

According to Jin et al., (2008), low-dose irradiation in the early phase may cause a temporary attenuation of fundamental cellular processes such as transcription and translation, along with an activation of cellular defense mechanisms associated with stress response. After 12 h of exposure, it is likely that cellular activities were resumed and defense systems were suppressed. Among genes in this cluster there are genes associated with cellular defense mechanisms in the early phase which included the TNF/stress related signaling pathway, the IL-5 signaling pathway, eosinophils in the chemokine network of allergy responses, the first multivalent nuclear factor signaling pathway and skeletal muscle hypertrophy regulated via the AKT/mTOR pathway (Jin et al., 2008).

Ionizing Radiation and Subcellular Structures

Extrachromosomal DNA is any DNA that is found outside of the nucleus of a cell. It is also referred to as extranuclear DNA or cytoplasmic DNA (Kuttler and Mai, 2007). Extrachromosomal DNA has been associated with genomic instability in eukaryotes (WPL, 2016). The sequences of cytoplasmic DNA are different from nuclear DNA in the same organism, showing that cytoplasmic DNAs are not simply fragments of nuclear DNA (Koch, Vogt and Kissel, 1983). In eukaryotes extrachromosomal DNA is primarily found in organelles (Rush and Misra, 1985). By its very nature, extrachromosomal DNA is structurally different from nuclear DNA. Cytoplasmic DNA is less methylated than nuclear DNA in the same organism (Koch, Vogt and Kissel, 1983). Extrachromosomal circular DNA (eccDNA) consists of repetitive sequences of DNA found in both coding and non-coding regions of chromosomes. EccDNA can vary in size from less than 2000 base pairs to more than 20,000 base pairs (WPL, 2016). Extrachromosomal circular DNA is present in all eukaryotic cells, and is usually derived from genomic DNA, both coding and non-coding regions of chromosomes (Cohen, Houben and Segal, 2008). The production of eccDNA elements from genomic DNA sequences adds to the plasticity of the eukaryotic genome and can influence genome stability, cell aging and the evolution of chromosomes (Cohen et al., 2010). Other extrachromosomal elements that are associated with genome instability are Double Minute Chromosomes (DMs), present in cancer cells (WPL, 2016). DMs are thought to be produced through breakages in chromosomes or overreplication of DNA in an organism (WPL, 2016).

Mitochondrial DNA is a main source of extrachromosomal DNA (WPL, 2016). Exposure to substances that affect the mitochondria, increasing the amount of oxidative damage to proteins is unique (Lin and Beal, 2006; Reynaud, 2010). It is well known that point mutations and alternative gene splicing of mtDNA have been linked to diseases that affect the heart, central nervous system, endocrine system, gastrointestinal tract, eye, and kidney (Chinnery and Turnbull, 1999).

The protein activities are compromised with the increasing LET. It has been clarified that alpha-irradiation causes great loss of the enzymatic activity.

The mean free path of alpha particles from Pu is very short, less than a few micrometers in living material. Thus it could cause much serious DNA damage rather than ^{137} Cs or ^{131}I. ROS and/or wave propagation can alter the cell membrane, as well as the nuclear and mitochondrial membranes, or endomembrane system of endoplasmic reticulum and the Golgi (Pouget et al., 2008).

Nuclear membrane is essential to enclose displaced chromosomes or chromosome fragments. Haaf and coworkers (1999) reported that MN originated from chromosomal material that was not incorporated into daughter nuclei during cell division. Different MN types depend on different agents. MN caused by ionizing radiation or clastogens mostly contain acentric chromosome fragments. Determination of MN frequencies has been widely used as a dosimeter of human exposure to radiation or clastogenic and aneugenic chemicals, and for the detection and risk assessment of environmental mutagens and carcinogens. The authors stressed that in spite of the utility of the MN test as an in situ monitor of cytogenetic effects, our understanding of the connection between initial DNA damage and the formation of MN was still poor. More recently, Fenech and coworkers (2011) explained the concept which implied that high MN frequency was related to cancer.

The increased demand for the secretory machinery in the endocrine cells, as indicated above, may result in the development of endoplasmic reticulum stress. Proteins destined for exocytosis are transported through the ER and *trans*-Golgi complex and stored in secretory granules prior to secretion. Only correctly folded proteins are transported to the Golgi compartment (Westermark, Andersson and Westermark, 2011). Ionizing radiation can trigger ER stress and cause an obstruction of the ER-Golgi transport. According to Westermark, Andersson and Westermark, (2011), an accumulation of unfolded proteins triggers upregulation of ER-located chaperones to assist folding of aggregation-prone proteins. Accumulation of misfolded proteins can cause numerous degenerative diseases. Misfolded proteins aggregate into amyloid. The most frequent neurodegenerative diseases caused by misfolding, such as Alzheimer's disease, Parkinson's disease, Huntington's disease, or no-neurodegenerative ones, like Diabetes Mellitus, have similar amyloidal origins. Many environmental factors can increase the risk of degenerative diseases onset (Reynaud, 2010).

Exposure to external or internal, natural and man-made radiation sources, induces oxidative damage of the cell, which causes structural disorganization of the cell membrane and the endomembrane system. As a consequence, protein misfolding ensues, which plays a key role in the pathogenesis of numerous diseases (Zunic, 2013-1; Zunic and Rakic, 2013).

Radioadaptive Mechanisms

According to a study authored by Matsumoto, Takahashi and Ohnishi, (2004), radiation-induced adaptive responses were observed in vitro and in vivo. The dose required to induce effective adaptive responses is within the range 0.01– 0.2 Gy, with low LET radiation. When dose is over 0.2 Gy, adaptive responses are scarcely induced, and when it is over 0.5 Gy, adaptive responses are almost never induced.

Experimental data showed that the lowest doses did not trigger the repair mechanism of the cells (Burlakova, 1999). Direct exposure of cells to a low dose of ionizing radiation can induce a condition of enhanced radioresistance, i.e., a "radioadaptive" response. Radioadaptive response could be understood as a kind of BSE, most likely induced by paracrine regulation by growth-promoting activity of the cells exposed to a low dose of alpha radiation (Iyer and Lehnert, 2002). The DNA damage-control biosystem is physiologically operative on both, metabolic and radiation induced damage, both affected predominantly by free radicals (Pollycove and Paperiello, 1997). The authors concluded that adaptive responses were suppressed by high-dose and stimulated by low-dose radiation. The potent cellular repair mechanisms are stimulated after radiation induced damage in case of low doses ... "Whatever the mechanisms, they seem able to act not only on the lesions induced by ionizing radiation but also on at least a portion of the lesions induced by some other toxic agents" (Pollycove and Paperiello, 1997). Other authors reported that effects of electromagnetic radiation (EMR) on the cell are much more complex: "The nature of EMR interaction with biological objects is determined by both parameters of radiation (frequency or wavelength, propagation speed, coherence of oscillations, and polarization) and the physical properties of a biological object as a medium in which an electromagnetic wave propagates (dielectric constant, electric conductivity, and nonlinear-dynamic parameters of information biopolymers, as well as parameters depending upon this quantities, namely, wavelength inside the tissue, penetration depth, and reflectance at the air–tissue interface)" (Reshetnyak et al., 1996).

There are many theoretical approaches used to explain the interaction between EMR and the living cell. These interactions may occur primarily in the plasma membrane, and then a cascade of changes propagates from the membrane to the nucleus of the cell (Behari, 1999). The magnetic field itself may interact with cellular components (Inan, 2005), affecting many targets in the cell, including cell membrane, DNA, metabolic and signaling pathways. The permeability properties of cell membranes can be affected by protein clusterization. Low energy EMF induces the stress response directly interacting with DNA. Increasing EMF energy in the radiofrequency range can lead to breaks in DNA strands. An EMF reactive DNA sequence, which acts as an EMF sensitive antenna, can be transferred to other gene promoters (Blank and Goodman, 2009). Many of these changes can be reverted after EM radiation is stopped. If changes could not be reverted, then the increased proliferation of cells may lead to carcinogenesis (Marinelli proceedings, 2007).

Chapter 6

HEALTH EFFECTS OF DEPLETED URANIUM

Inhalation of aerosols contaminated with the fumes of depleted uranium oxides is the dominant path of contamination. Internally deposed nano- or micro-sized particles of DU induce the low-slow dose effects in living tissues. We have already discussed a reciprocal relationship between low-slow doses of ionizing radiation and their harmful effects on the cell. Biological effects are extensive and unpredictable, owing to their complex regulation, dependent on exposition, as well as on genetic predisposition, health, age and other individual factors. Due to long pulmonary retention of 1,470 days, as expected in the case of inhalation of uranium oxides (Durakovic, 1999), a wide range of clinical manifestations can occur, depending on the individual predispositions of the exposed persons.

Radioadaptive tissue response of pulmonary tissue was described in the publications authored by Zunic (2013; 2013-1), in the context of the early, as well as delayed health effects of depleted uranium (Zunic, 2015). The early health effects of DU exposure are discussed in relation to the cytological properties of lung washings from children from the territory where DU ammunition was used. The importance of apoptosis has been stressed out, as well as the cytological profile of lung washings in the case of low-dose radiation, which is imperative in understanding the difference between alveolitis induced by low-dose radiation and that induced by high radiation doses. A recent publication proposed the hormesis-threshold model in genesis of lung cancer in smokers, who were exposed to internal sources of low-dose radiation originating from tobacco smoke, as well as from the environment. In both cases an exact measuring of tissue doses was not required. These models make evaluation of the health effects of internally deposed alpha-emitting radionuclides, including DU, easier (Zunic, 2015). This method enables a quick orientation concerning the extent of damage of complex tissue regulatory mechanisms in situ. Indirectly, it may indicate the existence of adaptive tissue response, premalignant or malignant lesions, which were considered the delayed health effects of DU.

The overwhelming radioadaptive/radioprotective tissue capacity may later cause autoinflammatory/autoimmune disorders accompanied by numerous symptoms and degenerative and inflammatory diseases. All tissues with oxidative metabolism were targeted, particularly kidney and bone. Cancer is one of the late consequences of low-dose radiation exposure (Zunic, 2013-1).

The recent war history of the Balkans and the Persian Gulf in the period 1992-2002 relates to many medical phenomena that can be observed in the light of present repeated exposure to low doses of radiation originating from the military use of depleted uranium.

Gulf Syndrome can be understood as a disease caused by repeated exposure to low doses of mixed (α, β, γ), predominantly alpha radiation originated from DU decay. Neurological/psychiatric manifestations of disease are in part the consequence of demyelination and together with other diseases with multisystem involvement are not rare. Psychiatric disorders are more frequent in the Gulf War veterans than in any other population, but without precise cause and effect relationship (Li et al, 2011). Gulf/Balkan War Syndrome may not be exclusively a disease of soldiers who participated in these wars. Taking into account prolonged exposures to alpha radiation (from the blast and later due to corrosion of armaments), Gulf/Balkan syndrome can be understood as a multicausal disease with multisystem involvement and time-dependent expression of symptoms from no cancerous diseases, to cancers in later phases affecting soldiers, as well as overall civilian population. Further details on this hypothesis are presented in a conceptual map in Figure 9 (according to Zunic, 2013-1). The arrows show the radiotoxic effects of DU on the cell. We suppose that following the deposition of DU particles in thoracic lymph nodes and bearing in mind long half-life of DU decay members, alpha particles induced the bystander effect and influenced the immunocompetence of T cells. Aggressive T cells may be responsible for delayed effects of alpha radiation. Repeated exposures to low/slow radiation doses may induce the LEC-phenomenon in the BAL specimens. Gulf/Balkan syndrome develops as a multicausal disease with multisystem involvement and time-dependent expression of symptoms, from non-cancerous diseases, to cancers in later phases.

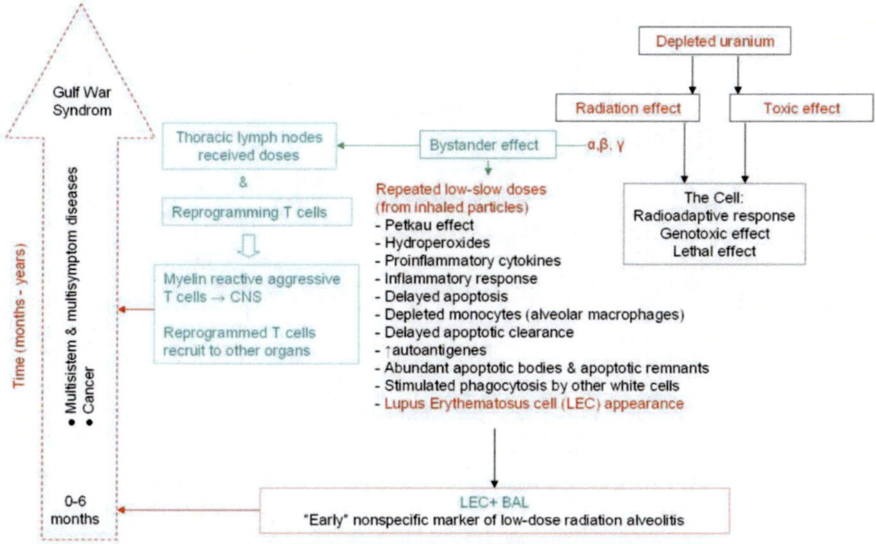

Figure 9. A conceptual map on early and delayed health effects of depleted uranium. Adapted with permission from primary data of Zunic, (2013-1).

Illustration of Early Health Effects of DU - Cytological Characteristics of Pediatric Lung Washings from the Territory Bombed with DU Projectiles

The study is based on authentic medical observation of differential cell counts of broncholaveolar lavage samples from 225 pediatric patients. All the patients, whose bronchoalveolar lavage samples were analyzed, originated from the territories which were geographically close to each other (Serbia and Montenegro seaside and Bosnia and Herzegovina, the territories of the former Yugoslavia (Figure 10) that were repetitively stroked by DU armaments).

After the bombing with DU projectiles, widespread air pollution due to the release of radioactive particles, both in the explosion, and in later years due to corrosion of missiles, could be expected.

Figure 10. Schematic representation of territorial origin of randomly selected 20 pediatric patients who underwent bronchoscopy examination after the bombing of Serbia 1999. The distribution of patients from the territory of the former Yugoslavia with: Lupus Erythematosus Cell-negative bronchoalveolar lavage specimens (○); Lupus Erythematosus Cell-positive bronchoalveolar lavage specimens (●). Reproduced with permission from primary data of Zunic (2013).

The BAL specimens were obtained from unrelated children who underwent flexible bronchoscopy for clinical symptoms and radiological changes consistent with the persistent pulmonary infiltrates at The Department of Pulmonology, Mother and Child Institute in Belgrade. They showed no clinical and/or laboratory signs of systemic Lupus Erythematosus or any other connective tissue disease (Zunic et al., 1996). BAL differential cell counting was performed in the laboratories of the Clinical Center of Serbia, in the period December 1992 – December 2002.

The laboratory protocol for cytological analysis of cytocentrifuge preparations of the BAL samples was applied (Haslam et al., 1984) with modification, as Hemaccel (Jugoremedia) was used as a medium for preparation of cell suspensions (Zunic's Doctoral Thesis, 1993). The samples were spun down freshly and pellet was adjusted with Hemaccel and cytocentrifuged, then air-dried and stained with May-Grünwald-Giemsa for differential cell counting. Slides were analyzed on an OPTON Axioplan light microscope for at least 500 cells.

The differential cell counting of pediatric BAL specimens, revealed cytomorphological feature of Lupus erythematosus cells (LEC) (Figure 11) in 47 cases of which not a single case was associated with coexistent autoimmune disease (Table 4).

The BAL cytocentrifuge preparations were examined by light microscopy and a number of cells showing pathognomonic cytological features of Lupus Erythematosus Cell was identified (Figure 11a, b). There is a LEC in the centre surrounded by other BAL cells, predominantly neutrophils (Figure 11a). LE cells were enlarged (25-30 μm in diameter), round shaped, with a peripheral, elongated nucleus and abundant cytoplasm containing a round to oval, well-circumscribed inclusion body of homogenous, basophilic appearance (Figure 11 b). More details and explanation on the LEC phenomenon in bronchoalveolar lavage of patients of whom not a single case was associated with coexistent autoimmune disease at the moment of trial were described by Zunic and coworkers in 1996.

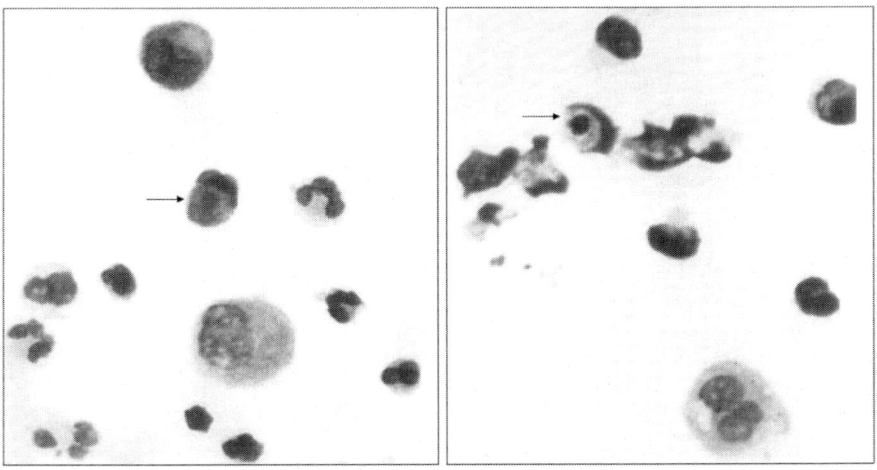

a. b.

Figure 11. Microphotograph of Lupus Erythematosus like cells in the bronchoalveolar lavage specimen of the patient with thalassemia minor combined with hypogammaglobulinemia and suppurative lung disease. Lupus Erythematosus like cells (arrows) in two microscopy fields of the bronchoalveolar lavage specimen (May-Grunwald-Giemsa, original magnification x330); a. Lupus Erythematosus like cell in the centre surrounded by other BAL cells; b. Intracytoplasmic Lupus Erythematosus body in enlarged cell with peripherally located binucleated nucleus. Reproduced with permission from primary data of Zunic (2013-1).

The results were issued consecutively, but after a full analysis classified into five time intervals (Table 4, Table 5 and Figure 12):

1) 1992-1993 – the time after the First Gulf War (1990–1991) - (A),
2) 1994-1995 – the time during the conflict and air strikes against targets in Bosnia - (B),
3) 1996 -1999 (March 24th) – the time between the bombing of targets in Bosnia and the bombing of targets in Serbia - (C),
4) 1999 (July-December) - the time after air strikes against targets in Serbia - (D),
5) 2000-2002– (E).

Table 4. Presentation of the numbers of the BAL samples analyzed, the LEC positive BAL samples and relative percentages of LEC in the BAL differential cell counts for individual specimens per year in the observed period 1992-2002

Years	1992	1993	1994	1995	1996	1997	1998	1999	2000	2001	2002
Number of BAL samples analyzed	1	40	36	42	27	19	-	10	29	9	12
Number of LEC positive BAL samples	1	6	3	2	6	7	-	7	6	3	6
Percentages of LEC in individual BAL differential cell counts	0.30	1.30 0.30 2.00 0.98 0.30 0.30	1.00 0.70 0.70	0.70 1.30	0.30 0.70 1.90 2.20 0.30 1.30	0.30 0.90 0.30 0.30 3.60 0.30 1.00	-	0.69 0.40 4.50 8.40 4.70 1.50 3.90	0.20 0.20 1.20 0.60 0.30 0.60	0.60 1.40 0.30	3.43 1.14 0.30 0.60 0.30 0.30

Adapted with permission from primary data of Zunic, (2013-1).

Presented results (Tables 4 and 5) imply that there was higher percentage of the LEC positive BAL specimens, and an increase in the yield of LEC in differential cell counts in individual LEC positive BAL specimens during the D time interval (1999, July-December).

Statistical significance of differences of mean percentages of LEC in differential cell counts of LEC positive BAL specimens was observed and results presented in Table 6.

Table 5. Average percentages of the LEC positive BAL samples analyzed in time intervals corresponding to the period after the First Gulf War, during and after the bombing of Bosnia and Herzegovina and Serbia

Intervals of time	Number of BAL specimens analyzed	Average percentages of LEC positive BAL specimens
A. 1992-1993	41	17.07
B. 1994-1995	78	6.41
C. 1996 -1999 (until March 24th)	48	27.08
D. 1999 (July-December)	7	87.50
E. 2000-2002	50	30.00

Adapted with permission from primary data of Zunic, (2013-1).

Table 6. Statistical significance of differences between LEC percentages in differential cell count of LEC-positive and LEC-negative BAL specimens within observed intervals of time

Groups compared	A+B+C vs. D	A+B+C vs. E	D vs. E
t-test value	4.54	1.07	3.45
DF	31	38	21
Two tailed probability	0.0001	0.2906	0.0024
Statistical Significance	$p<0.001$	no significant	$p<0.01$

DF - degrees of freedom; estimated intervals of time: A. 1992-1993- the time after the First Gulf War (1990–1991); B. 1994-1995- the time during the conflict and air strikes against targets in Bosnia and Herzegovina; C. 1996 -1999 (until March 24th 1999) - the time between the bombing of targets in Bosnia and Herzegovina and the bombing of targets in Serbia (March 24th – June 10th 1999); D. 1999 (July-December) - the time after air strikes against targets in Serbia; E. 2000-2002 - the time after the bombing of Serbia and the end of the observed period. Adapted with permission from primary data of Zunic, (2013-1).

There is 1) a statistically highly significant increase of percentage of LEC in the time after air strikes against targets in Serbia (D, July-December 1999), when compared with the percentage of LEC in the period 1992-March 24th 1999 (A+B+C); 2) a statistically significant decrease of percentage of LEC in the time after air strikes against targets in Serbia (E, 2000-2002), in comparison with the first six months after the bombing of Serbia (D). No significant differences were found between mean percentages of LEC in the time after air strikes on targets in Serbia (E, 2000-2002) in comparison with the period preceding the bombing of Serbia – 1992 to March 24th 1999 (A+B+C) (Zunic, 2013-1).

Schematic presentation of the percentages of (LEC+) the BAL specimens in the observed time intervals was shown in Figure 12.

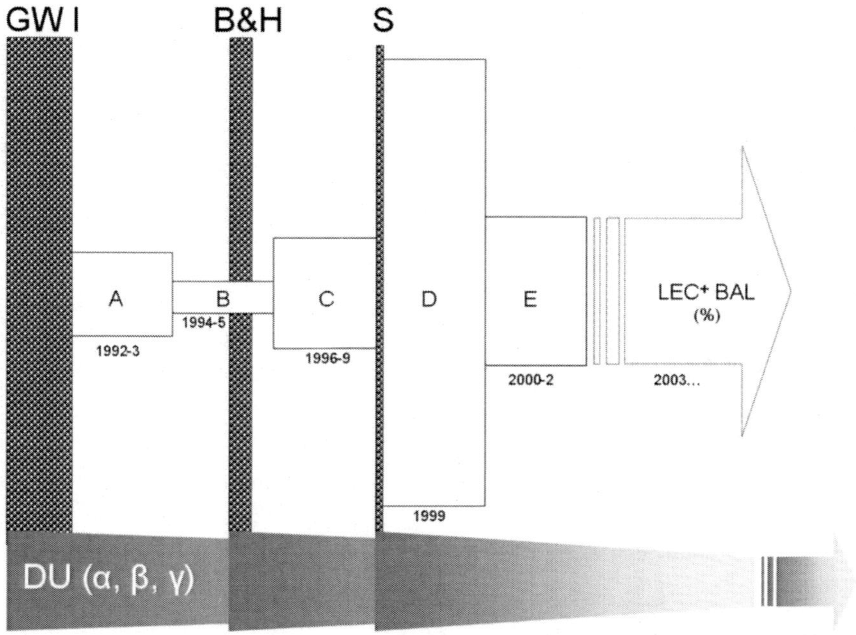

Figure 12. Schematic presentation of a time-dependent appearance of LEC in relation with the use of projectiles with depleted uranium (DU) and proposed α, β, γ-radiation exposure. Black&white areas – the symbolic presentation of missiles with depleted uranium used in GW (The First Gulf War), B&H (Bosnia and Herzegovina) and S (Serbia), the republics of the former Yugoslavia, with time correlates; the white area squares represent relative percentages of the Lupus Erythematosus Cell positive bronchoalveolar lavage specimens in five time intervals (A-E): Presentation of average percentages of the Lupus Erythematosus Cell positive bronchoalveolar lavage specimens in predefined time intervals: A. 1992-1993- the time after the First Gulf War (1990–1991); B. 1994-1995- the time during the conflict and air strikes against targets in Bosnia and Herzegovina; C. 1996 -1999 (until March 24th 1999) - the time between the air strikes against targets in Bosnia and Herzegovina and bombing of targets in Serbia (March 24th – June 10th 1999); D. 1999 (July-December) - the time after the air strikes against targets in Serbia; E. 2000-2002 - the time after the air strikes against Serbia and the end of the observed period; the grey arrows (bottom of Figure 12) represent proposed exposure to radiation from depleted uranium radioactive decay, with time-dependent attenuation (dark to pale gray arrows) and with hypothetical amplification, simultaneously with nuclear ammunition repeated use. Adapted with permission from primary data of Zunic, (2013-1).

The time and place of the detached use of munitions containing DU may be connected to an implicit dynamics of radioactive-particles exposure, which has been prolonged, and has not ended with the end of air strikes. Each new air strike caused a new increase in DU-particles exposure, which is presented schematically at the bottom of figure 12 (Zunic, 2013-1).

With the aim to better characterize differences of (LEC+) and (LEC-) alveolitis, a comparative and correlation analysis has been performed in 20 BAL specimens (Zunic, 2013). The results of differential cell counting of lung washings of patients from (LEC+) group and (LEC-) group, with respect of neutrophils (N), eosinophils (E), lymphocytes (Ly), mast cells (Mast), alveolar macrophages (AM), Lupus Erythematosus Cells (LEC) and other cells (related to the sum of epithelial cells, plasmocytes and monocytes in the BAL specimens) were compared (Table 7). A significant increase of percentages of neutrophils and eosinophils and decreased percentages of macrophages were found in (LEC+) BAL specimens in comparison with (LEC-) ones ($p<0.05$, $p< 0.001$, $p< 0.001$, respectively), according Zunic, (2013).

Table 7. Statistical significance of differences for differential cell counts between the LEC-positive and LEC-negative BAL samples

	LEC- Mean & SD	LEC+ Mean & SD	Significance of differences
Neutrophils	7.09 ± 2.76	21.42 ± 15.89	p<0.05
Eosinophils	8.07 ± 5.42	26.68 ± 9.90	p<0.001
Lymphocytes	15.24 ± 8.55	13.35 ± 6.19	no significant
Alveolar macrophages	59.33 ± 14.47	27.57 ± 18.13	p<0.001
Mast cells	0.48 ± 0.32	0.63 ± 0.43	no significant

Adapted with permission from primary data of Zunic, (2013).

The results of correlation analysis of BAL cell populations in (LEC+) vs. (LEC-) specimens are presented in Table 8. Statistically significant negative correlation was found between BAL eosinophils and alveolar macrophages in (LEC+) BAL specimens; (LEC-) BAL specimens showed statistically significant negative correlations between neutrophils' and lymphocytes' relative percentages.

Table 8. Correlation analysis of BAL cell populations in LEC- positive vs. LEC- negative specimens

Correlations	LEC-	LEC+
N/Ly	r = -0.75, p<0.05	no significant
E/AM	no significant	r = -0.75, p < 0.05
Ly/AM	no significant	no significant
AM/Mast	no significant	no significant

Adapted with permission from primary data of Zunic, (2013).

The origin of LEC in BAL has been understood as a possible effect of DU fumes exposure. All patients included in this study, originated from different polluted areas in the territory of the former Yugoslavia and all of them were in the regions of air pollution after the First Gulf War (Figure 1) (Zunic, 2013; Zunic, 2013-1).

The presence and a significant increase in frequency of the (LEC+) BAL samples in examinees during the period 1992-2002, detected after the NATO bombing of Serbia, enabled search for the relationship between the presence of LEC in BAL and DU originated alpha radiation. The increased yield of (LEC+) BAL samples after the bombing of Serbia (March 24 to June 10, 1999), and a rough estimation on the "unusually" high succumb of LEC in individual BAL specimens, were found in the first 6 months after air strikes against targets in Serbia. The results presented the clearly implicate time-dependent effects of DU (Zunic, 2013; 2013-1).

The publications authored by Zunic (2013; 2013-1) opened some questions related to the extensiveness of the α-particle-induced bystander effect, observed early, and discussed the delayed health effects of depleted uranium.

Medical entities, as Balkan's or Gulf War Syndrome, are subsumed into a common entity whose pathogenesis is based on radiobiology of the time-dependent effects of DU in military personnel, as well as exposed civilians. Finally, these studies highlighted the LEC phenomenon in BAL as an early radioadaptive tissue response and one of the main, although non-specific characteristic of DU induced low-dose radiation alveolitis.

This is essential for detecting the differences between low-dose and high-dose radiation pneumonitis, which may be induced accidentally, or by therapeutic procedures.

Toma et al., (2010) reported lymphocytic alveolitis as a marker of radiation pneumonitis in patients after radiotherapy for breast cancer. This high-dose therapeutic irradiation induced pneumonitis, by its pathogenesis, is different from DU low-slow-repeatedly doses induced pneumonitis. In (LEC+) BAL specimens predominates neutrophilic/eosinophilic alveolitis with a marked decrease of AM (Table 7).

It was published earlier that radiation affects the blood cells, thus leading to a dominant decrease in the number of monocytes (IAEA, 1997). Having in mind that alveolar macrophages are tissue counterparts of circulating monocytes, the results support an idea that (LEC+) BAL specimens may reflect radiation induced changes.

Neutrophils are not only major effectors of acute inflammation, but they can also contribute to chronic inflammatory conditions and adaptive immune responses (Kolaczkowska and Kubes, 2013). With a deeper look into the mechanism of pathogenesis of (LEC+) alveolitis, based on the results in Tables 7 and 8, it is very likely that a massive migration of neutrophils from the circulation to the pulmonary tissue takes place at the expense of lymphocytes, which also, but to a lesser extent migrate from the bloodstream to the pulmonary tissue. There is a possibility of reciprocity in the migratory potential of different cell populations from the circulation into the tissue (namely, eosinophils, or monocytes that mature into the AM). A slight increase of mast cells (Table 7) may reflect the aggressive pathogenic properties of (LEC+) alveolitis, probably regulated by cytokines and deranged intercellular interactions (Zunic, 2013).

The conventional understanding of the nature and pathogenesis of LEC appearance in the body fluids and tissues has been in relation with coexistent autoimmune diseases, particularly Systemic Lupus Erythematosus (Hepburn, 2001; Gulhane and Gangane, 2012). A recent paper pointed out that the presence of LEC in BAL could be a manifestation of early radioadaptive/radioprotective tissue response to exposure to low doses of radiation, particularly alpha particles that were trapped in the airways (Zunic, 2013-1). The author's observations since 1996 (Zunic et al., 1996; Zunic, 2013; Zunic, 2013-1) positioned the presence of LEC in BAL as a nonspecific in pneumopathies that occurred coincidentally with the radiation exposure in the territory that was stroked by missiles with DU.

These statements, together with the Petkau effect interpretation (Figure 5), clearly indicate the missing links in the pathogenesis of the LEC phenomenon. The low-slow-dose conditions lead to dramatic depletion of monocytes, failure of the lymphocyte immune system, and consequently, depressed cellular immune system (IAEA, 1997; Bertell, 2006). Otherwise, alteration of monocytes, as circulating precursors of tissue macrophages, changes subsequently immunocompetence and the phagocytic properties of alveolar macrophages, originating from recruited monocytes. Owing to radiation and alpha particle-induced bystander effect, massive apoptosis can be achieved. Defective apoptotic clearance, due to impaired phagocytosis by macrophages, may induce overexpressed phagocytosis by other cells of white lineage with concomitant LEC appearance (Cacciapaglia et al, 2009). There is a relationship between delayed apoptotic cell clearance and lupus-like phenotype (Cohen et al, 2002). Apoptosis also plays a role in the regulation of T-cell populations (Lohman and Welsh, 1998; Stuart and Hughes, 2002).

An autoimmune (auto-inflammatory) response induced by external agents was described as ASIA syndrome (Israeli, 2012). Detection of LEC in BAL is in favor of the multi-causal character of ASIA syndrome. Although occurrence of LEC in BAL has been described earlier as a nonspecific phenomenon (Zunic et al, 1996), the present study observed the appearance of LEC in BAL as a possible consequence of inhaled DU particles. Bearing this in mind, the LEC phenomenon in BAL can be understood as a nonspecific tissue response to a noxious agent leading to the altered properties of phagocytosis and apoptotic clearance. If radiation is a trigger, LEC appearance in BAL specimens may be understood as an early radioadaptive mechanism and a hallmark of radiation alveolitis caused by DU particles that are trapped in the airways. The LEC phenomenon in BAL may occur in an attempt of pulmonary tissues to remove debris through other white cell lineage under conditions of vulnerable phagocytosis by alveolar macrophages. Taking this mechanism into account, the LEC phenomenon may also be a radioprotective phenomenon in the conditions of lack of radioprotective mechanisms at low dose radiation (Zunic, 2013-1).

The changed migratory capacity of cells from the bloodstream to the pulmonary tissue, as well as the nature of cell-to-cell interactions may be the cause of key differences between (LEC+) and (LEC-) alveolitis. Very possibly, the recruitment of different cell populations from circulation into the tissue is stimulated by the same harmful agent, but cell-to-cell communication may be changed in conditions of simmering bronchopulmonary pathogenic process. At the same time, the tissue opposes long-lasting inflammation as well as diminished apoptotic clearance by alveolar macrophages, inducing adaptive mechanisms including the LEC phenotype expression (Zunic, 2013).

DU also crosses the placenta and is stored in the fetus (Shuryak, Sachs and Brenner, 2007). Lupus Erythematosus cells were found in two children who were lavaged in October and December 1999, 3 and 8 months old at the time of clinical trial respectively. The first child was born in the suburb of Kraljevo, one month after the end of the NATO bombing of Serbia, and the other one, in Mladenovac, a month after the start of the bombing of Serbia (Zunic's unpublished data related to laboratory evidence and medical documentation). It is very possible that their intrauterine exposure to DU particles was a possible trigger for later clinical pulmonary manifestations. This finding supported the understanding of radioprotective (rather than radioadaptive) response in the postnatal period (Zunic and Rakic, 2015).

Illustration of the Delayed Health Effects of DU - Cytological and Cytochemical Characteristics of Lung Washings of Smokers and Lung Cancer

As cigarette smoke affects the main lung defense system at both, morphological and cellular level, it was important to observe the dynamcs of the apoptotic bodies' generation process, originating from mononuclear cells and their removal by resident macrophages (Zunic et al., 2004).

The delayed apoptosis of AM represents one of the mechanisms for accumulation of these cells (Schaberg, 2000). Some in vitro studies suggested that exposure to smoke could induce the apoptosis of AM (Aoshiba, 2001). Since macrophages are interposed between stress, inflammation and immune response in the maintenance of basic survival response (Ottaviani and Franceschi, 1997), it is a question how they can participate in the process of apoptosis in tissues exposed to different pathogenic and cancerogenic stimuli. AM might be important in the resolution of acute inflammation by the ingestion of apoptotic neutrophils. Granulocyte apoptosis is important in the resolution of pulmonary inflammation. We described the semiquantitative cytochemical method to evaluate the contribution of AM in clearance of pulmonary tissue in situ during the process of programmed cell death. As the main pulmonary resident phagocytes, AM are actively involved in the process of apoptotic bodies removal. Apoptotic bodies are generated from non-resident immunocytes, recruited "on demand" from blood to the lung (Zunic et al., 2004; Zunic et al., 2007).

This discussion reconnects known facts about smoking exposure and lung cancer (Zunic et al., 2004; Zunic, 2007) with the existing low-level-radiation-doses from tobacco smoke, as well as from environmental radioactive air-pollutants. It is not possible to exclude the fact that all patients (smokers and nonsmokers alike) were exposed to low-dose DU radiation originating from inhaled pollutants after the bombing of FRY, or the Persian Gulf, according to our publications (Zunic, 2013; Zunic, 2013-1; Zunic and Rakic, 2015) and Figure 1. We considered nonsmokers as exposed only to environmental doses. Figure 13 is a schematic presentation of radionuclides from air pollution and tobacco smoke in the bronchoalveolar space of nonsmokers and smokers.

As reported by WSDH (2002), only 81% of annual dose is from natural sources. The rest of the dose comes from man-made background radiation. The average annual whole body dose is 361 mrem/year, plus additional ~280 mrem/year for smokers.

Alpha emitting radionuclides in tobacco smoke, or tissue embedded uranium particles which are mixed, alpha, beta, gamma emitters, maintain continuous very-low doses of irradiation of the surrounding tissue structures. Low doses and dose rates stimulate protective intercellular and intracellular signaling (Scott, 2008).

With increasing radiation dose and dose rate, DNA repair systems are less effective than at low doses or dose rates (Tubiana, 2009). Higher radiation doses are in relation to increased cancer risk (Scott and Di Palma, 2007). A publication by Tubiana et al., (2009) suggests the existence of real or practical thresholds for carcinogenesis. The protective mechanisms that lead to activated natural protection against cancer and other genomic-instability-associated diseases are understood as evolutionally highly conserved (Mitchel, 2006; 2007). Gene transcription studies revealed highly specific genes to ionizing radiation (Park et al., 2002; Roy, Gruel and Vaurijoux, 2009) and differences between low-dose and high-dose responses (Fachin et al., 2009).

A generalized response to mild stress, above an individual-specific threshold level, is associated with the protective signaling, which is estimated mostly in relation to apoptotic tissue response (Scott, 2005). As the universal ability of tissue, apoptosis can be influenced by broad spectrum of stressors, including radiation (Zunic and Rakic, 2015). Apoptosis is one of the earliest radioprotective mechanisms (Zunic, 2013). When adaptive mechanisms resisting smoking (& radiation) synergic influences become overwhelmed, a cancerous lesion may develop. We indicated the existence of the complex mechanisms of radiation hormesis that are integrated in the biosystem's output response to environmental inputs (Zunic and Rakic, 2013).

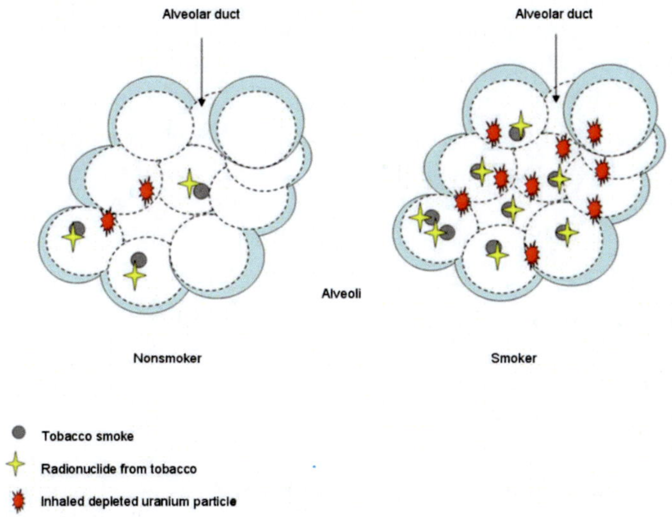

Figure 13. Schematic presentation of radionuclides from air pollution and tobacco smoke in the bronchoalveolar space of nonsmokers and smokers.

Given the different models of low-dose radiation risk assessment that were tested (Elgazar and Kazem, 2006), the aim of this study is to highlight a neural network model for dose-response relationship based on apoptotic parameters. We observe this graph, presented bellow, as a clinical nomogram suitable for further testing so as to determine whether every new examinee is at risk by belonging to one of the groups, representing healthy nonsmokers, control smokers and patients with lung cancer. This method is a simple and fast way to combine the advantage of neural networks in relation to statistical data processing in this specified case. We decided to present a methodological approach used for nomogram construction, according to Zunic et al., (2007).

Materials and Methods

This study is designed as a double blind study. All the patients underwent bronchological unit in order to perform routine diagnostic procedures. Bronchoalveolar lavage was performed in 22 examinees. After completing clinical and laboratory investigation (including histopathological analyses), it was possible to separate three groups: 9 healthy nonsmokers, 6 control smokers and 7 patients with non-small-cell lung cancer (NSCLC) (all of them were smokers). The patients were without any therapy at the moment of bronchological examination and the diagnosis of NSCLC was clinically and pathologically confirmed.

All the patients who were considered as the control groups underwent bronchological unit because of a longstanding cough or bleeding, but no signs of lung disease were found after a detailed clinical and bronchological examination.

Smoking Exposure

Total smoking exposure in the group of smokers was calculated as follows and presented in Table 9:

Pack-years = (Age of presentation-age started-years stopped) x Pack/Day

The smoking exposure values in control smokers were 1.5, 1.2, 1.5, 1, 0.6 and 5.0 and in smokers with NSCLC, the calculated values of smoking exposure were: 40, 55, 87.5, 25, 36, 35 and 30. Passive smoking was not considered for nonsmokers, while for all examinees total smoking exposure was 0 (Zunic et al., 2004; Zunic et al., 2007).

TUNEL Assay

Apoptotic detection in cytocentrifuge preparations of BAL cell suspensions was evaluated by light microscopy using TUNEL in situ cytochemical method (Boehringer Mannheim, In Situ Cell Death Detection Kit, POD; Cat. No: 1 684 817) (Milosevic, Rakic and Ruzdijic, 1998; Milosevic et al., 2000), and modified for BAL cytocentrifuge preparations (Zunic et al., 2004). In order to destroy endogenous peroxidase, selected cytocentrifuged BAL preparations were incubated with blocking solution (0.3% H_2O_2 in methanol) for 30 min at room temperature. After that, the slides were rinsed in PBS (pH 7.4), immersed in permeabilization solution (0.1% Triton X-100 in 0.1% sodium citrate) for 15 min at room temperature, and rinsed twice in PBS for 5 min. The samples treated with DNase I (100 μg/ml) for 15 min at room temperature prior to labeling, served as positive controls. The slides were labeled for 1 h at 37 °C in a dark, humidified chamber. The labeling solution (50 μl) contained calf thymus TdT (terminal deoxynucleotidyl transferase) and a nucleotide mixture in reaction buffer with modified nucleotides (fluorescein-dUTPs). The enzymatic reaction was terminated by immersing the slides in 3.0 M sodium chloride, and 30 mM sodium citrate solution (TB) for 15 min at room temperature. The slides were washed twice in PBS (pH 7.4), covered with 2% aqueous bovine serum albumin solution for 10 min at room temperature, rinsed in PBS and incubated with 50 μl peroxidase-conjugated anti-fluorescein antibodies (Converter-POD) in a dark, humidified chamber for 30-60 min at 37 °C. After three rinses with PBS, the samples were stained with diaminobenzidine (DAB) solution (6 mg in 10 ml Tris-HCl, pH 7.6) containing 0.03% H_2O_2, for about 10 min at room temperature. Finally, the sections were lightly counterstained with Harris' hematoxylin, dehydrated in alcohol and cleared in xylene. The slides were mounted with Permount (Fisher Chemical) and analyzed under a light microscope (ZEISS, Axioplan) using immersion objective.

Indexing and Scoring Method and Calculation of Apoptotic Capacity

The property of AM to engulf apoptotic cells was estimated by light microscopy, including 1,000 features per sample. These features are related to the subsequent steps (adsorption, internalization and digestion of apoptotic bodies by alveolar macrophages). Based on the yield of each one, indexing and scoring method was performed as presented schematically in Figure 14 (Zunic et al., 2004).

Figure 14 is a schematic representation of the unique features from photomicrographs of BAL cytocentrifuge preparations stained by TUNEL. The percentage of each stage was multiplied with the corresponding index (starting from 1 to 4; FAB are not indexed – their percent is expressed separately) and the sum calculated. Then, the final score for single slide calculated was within the theoretical range from 100 to 400. The score is a numerical equivalent of tissue ability for apoptosis and removal of apoptotic bodies, i.e., it represents the apoptotic capacity (AC) of the tissue (Zunic et al., 2004).

Apoptotic Index

In the same specimens stained by TUNEL, the apoptotic index (AI) was estimated and calculated as a relative per-cent of apoptotic cells in relation to all cells observed on the slide for individual patient.

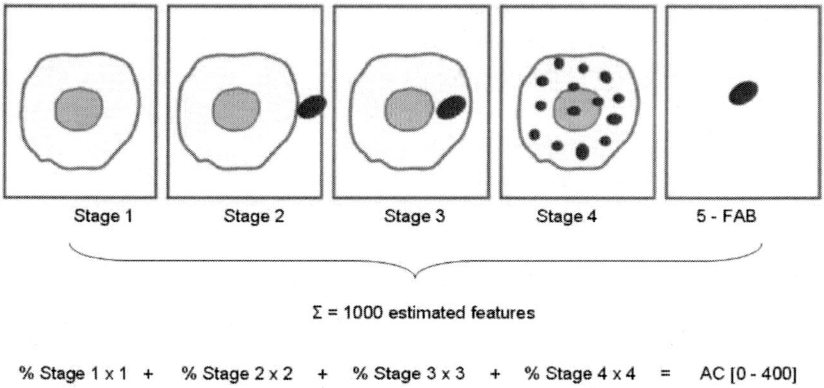

Figure 14. Schematic presentation of apoptotic capacity calculation. Schematic presentation of the features of interest and calculation of apoptotic capacity (AC): The big white nucleated cells represent AMs in four features important for the calculation and the small dark contours are free apoptotic bodies (FAB) (5): clear AM (1); FAB adherent to AM (2); FAB internalized by AM (3) and FAB digested inside AM (4). Each feature was expressed as a percentage of the sum: total macrophages (stages 1-4) + free apoptotic bodies = 1,000 features. Apoptotic capacity - a relation between the presence of free apoptotic bodies and their removal by phagocytosis was calculated as a sum of multiples of the corresponding stages (their relative percents) and weighing factors. Adapted from primary data of Zunic et al., (2004).

Statistical Analysis

A double tailed t-test was performed for estimation of the statistical significance of the parameters between the groups investigated.

The functional relationship between apoptotic capacity and apoptotic index is presented by a neural network method (from primary data of Minic, N.) (Zunic et al., 2007; Minic et al., 2011).

Results

The smoking exposure mean values and apoptotic parameters (AC and AI) in smokers and NSCLC are presented in Table 9. A review of smoking exposure values showed that all patients with NSCLC were heavy smokers (Table 9).

AI is gradually increasing in smokers and heavy smokers (the NSCLC group) in comparison with nonsmokers. AC is increased in smokers in comparison with nonsmokers, but in the NSCLC group, AC is significantly decreased both, in comparison with nonsmokers, as well as control smokers.

Table 9. Smoking exposure and apoptotic parameters in smokers and NSCLC

	Smoking exposure	AC	AI (%)	AC (compared to nonsmokers)	AI (compared to nonsmokers)
Nonsmokers	0	218.29±56.24	0.07±0.03		
Smokers	1.80±1.60	289.55±50.77	0.15±0.08	↑	↑
NSCLC	44.81±21.33	150.37±40.61	0.28±0.04	↓	↑
Nonsmokers vs. smokers		$p < 0.05$	$p < 0.05$		
Nonsmokers vs. NSCLC		$p < 0.05$	$p < 0.001$		
Smokers vs. NSCLC	$p < 0.001$	$p < 0.05$	$p < 0.01$		

Adapted with permission of Zunic et al., (2007).

Introduction to Neural Networks

A neural network is relatively new method in the area of artificial intelligence. It is used to simulate the human reasoning, especially in problems with pattern recognition. All the neurons inside the network are simplified to the mathematical level – potential natural impulses are expressed as corresponding numerical values. Similar to biological neuron, this artificial structure is consisted of its local memory and dendrites that connect it with other neurons. One of the dendrites can be considered as axon and it is responsible for output of the neuron, after the given information has been processed. Each connection has a certain weight that is multiplied with neuron data (signal) in order to adjust it to an acceptable level.

Cohonnen's model of neural network is based on a three-layer architecture, where each layer can have an infinite number of neurons. The input layer receptors are responsible for user to insert either new knowledge or question. The processing layer is used for learning and classification of new data, while the output layer effectors are used to express the final conclusion to the user. Each parameter is presented as one neuron in the input layer, so the complexity of input layer defines the dimension of the whole system. Problems with only two input parameters are solvable in two dimensions and can be clearly presented through the coordinate system. The main idea is to separate the plot into areas with common data inputs – responsibility of the processing layer. After the information is classified, it is easy to deduce the set of rules that will help in bringing up further conclusions. If there are more than one area with the same type of input, the problem is not linearly separable, so the processing layer will need one more sub-layer in order to classify data correctly.

Proposed Solution

In the present study, we estimated the relation between apoptotic parameters: apoptotic capacity and apoptotic index. Apoptotic index is expressed as a ratio yielding by apoptotic cells and apoptotic capacity reflects some dynamism in generation (presence) of apoptotic bodies and their removal by alveolar macrophages' phagocytosis. After evaluation of these parameters in the investigated groups (nonsmokers, smokers and NSCLC), the nomogram (presented in Figure 15) was carried out by a neural network method, by which is possible to predict belongings of every next patient to one of the groups, in regard of AC and AI analysis.

There are 22 patients (9 nonsmokers, 6 smokers and 7 NSCLC) included in this research, but one smoker does not have measured apoptotic index. That is why the network is built only on the sample of 21 patients. The data structure of this certain neural network is presented bellow in Table 10.

Table 10. Design of Neural Network

Case	AC	AI	TH1	TH2	TH3	Z1	Z2	Z3	TH4	Z4	Result
1	117.6	0.122	-0.02	170.40	-15.97	0	1	0	-0.50	0	Non-Smoker
2	212.7	0.110	-0.12	75.30	-121.07	0	1	0	-0.50	0	Non-Smoker
3	215.4	0.037	-0.70	72.60	-184.58	0	1	0	-0.50	0	Non-Smoker
4	260.1	0.040	-0.68	27.90	-226.78	0	1	0	-0.50	0	Non-Smoker
5	245.6	0.094	-0.25	42.40	-167.30	0	1	0	-0.50	0	Non-Smoker
6	285.2	0.092	-0.26	2.80	-208.56	0	1	0	-0.50	0	Non-Smoker
7	280.5	0.041	-0.67	7.50	-246.35	0	1	0	-0.50	0	Non-Smoker
8	161.2	0.042	-0.66	126.80	-126.21	0	1	0	-0.50	0	Non-Smoker
9	186.3	0.065	-0.48	101.70	-132.16	0	1	0	-0.50	0	Non-Smoker
10	291.4	0.079	-0.37	-3.40	-225.59	0	0	0	0.50	1	Smoker
11	360.8	0.054	-0.57	-72.80	-315.82	0	0	0	0.50	1	Smoker
12	260.7	0.149	0.19	27.30	-136.58	1	1	0	0.50	1	Smoker
13	238.9	0.174	0.39	49.10	-93.96	1	1	0	0.50	1	Smoker
14	245.9	0.276	1.21	42.10	-15.99	1	1	0	0.50	1	Smoker
15	230.0	0.340	1.72	58.00	53.22	1	1	1	0.50	1	NSCLC
16	143.9	0.304	1.43	144.10	109.33	1	1	1	0.50	1	NSCLC
17	134.8	0.235	0.88	153.20	60.96	1	1	1	0.50	1	NSCLC
18	168.2	0.258	1.06	119.80	46.71	1	1	1	0.50	1	NSCLC
19	106.2	0.286	1.29	181.80	132.04	1	1	1	0.50	1	NSCLC
20	150.8	0.200	0.60	137.20	15.80	1	1	1	0.50	1	NSCLC
21	118.7	0.310	1.48	169.30	139.53	1	1	1	0.50	1	NSCLC

In this moment the only columns in focus should be apoptotic capacity, apoptotic index and result. If these values are put in coordinate system, the chart will reach the following form (AC – X axes, AI – Y axes):

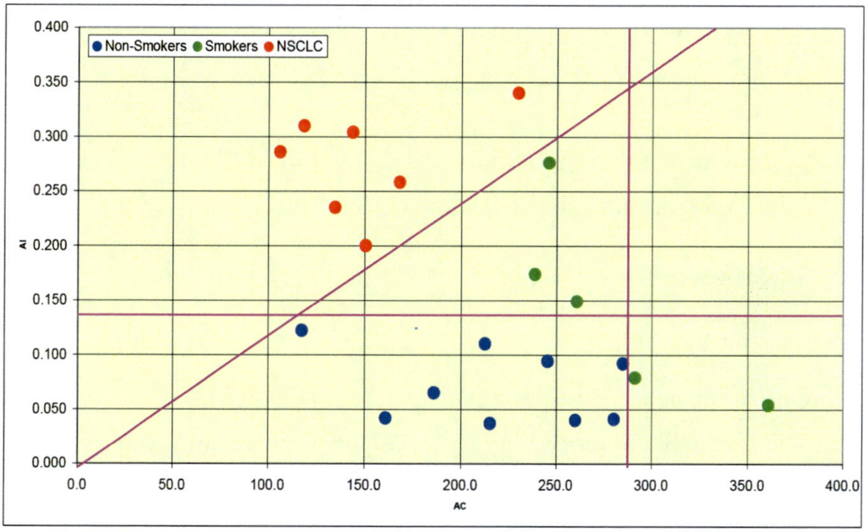

After separating the plain with three lines, it is obvious that the problem is not linearly separable, because there are two zones with smokers that could not be unified into one, at least not under this network design. The first step in design of the network is to estimate the formulas for given lines. It is well-known that universal formula is $ax + by + c = 0$.

Line 1 (containing points (0,0.125) and (1,0.125)):

$$a \cdot 0 + b \cdot 0.125 + c = 0 \wedge a \cdot 1 + b \cdot 0.125 + c = 0$$
$$\Rightarrow b \cdot 0.125 + c = 0 \wedge a \cdot 1 + b \cdot 0.125 + c = 0$$
$$\Rightarrow a = 0 \wedge b = -8c$$

From now on, a will be replaced with w_{n1}, b with w_{n2} and finally c with w_0, where n is the number of the certain line (neuron in processing layer).

Thus, one of the possible triplets is $(w_{11}, w_{12}, w_{10}) = (0,8,-1)$.

Line 2 (containing points $(288,0)$ and $(288,1)$):

$$w_{21} \cdot 288 + w_{22} \cdot 0 + w_{20} = 0 \wedge w_{21} \cdot 288 + w_{22} \cdot 0 + w_{20} = 0$$
$$\Rightarrow w_{22} \cdot 288 + w_{20} = 0 \wedge w_{21} \cdot 1 + w_{22} \cdot 288 + w_{20} = 0$$
$$\Rightarrow w_{22} = 0 \wedge w_{20} = -288 w_{21}$$

One of the possible triplets is $(w_{21}, w_{22}, w_{20}) = (-1,0,288)$.

Line 3 (containing points (0,0) and (250,0.3)):

$$w_{31} \cdot 0 + w_{32} \cdot 0 + w_{30} = 0 \wedge w_{31} \cdot 250 + w_{32} \cdot 0.3 + w_{30} = 0$$
$$\Rightarrow w_{30} = 0 \wedge w_{32} = -833 w_{31}$$

One of the possible triplets is $(w_{31}, w_{32}, w_{30}) = (-1,833,0)$.

The final table of processing layer weights is as the following:

$W_{11} = 0$	$W_{21} = -1$	$W_{31} = -1$
$W_{12} = 8$	$W_{22} = 0$	$W_{32} = 833$
$W_{10} = -1$	$W_{20} = 288$	$W_{30} = 0$

It is already explained that the first number in weight index presents the ID of the certain neuron in processing layer. The second number is the ID of the origin neuron from input layer which passed the data. Zero presents the constant which is not passed from any previous neuron and it is the unique measure how the certain neuron affects the signal.

Now Table 10 looks much more understandable. The TH values are calculated according to the following formula:

$$TH_{iC} = AC_C \cdot W_{i1} + AI_C \cdot W_{i2} + W_{i0},$$

Where C is the case number and $i \in \{1,2,3\}$.

The calculated weights precisely separated for every data input is it greater or less than each of the given lines. After the previous step of TH calculation, for every data input there are some new values that are greater or less than zero (depending on case being above or under the line). If the network threshold is defined at point of 0.5, all the cases are clearly separated in zeros (greater than 0.5) and ones (less than 0.5). That is how the Z values (Table 10) are estimated. From this point, it is obvious that there are certain patterns for each group – nonsmokers have Z values 0-1-0, smokers 0-0-0 and 1-1-0, while NSCLC have the value of 1-1-1. There is a small intersection between smokers' and nonsmokers' classes that demands another, more precise division. Taking into consideration that Z2 and Z3 values are the same for all smokers and nonsmokers, there is no need to view this problem in three dimensions, but just in two, for first two Z values. The following chart presents nonsmokers ($(0,1)$) and smokers ($(0,0)$ and $(1,1)$):

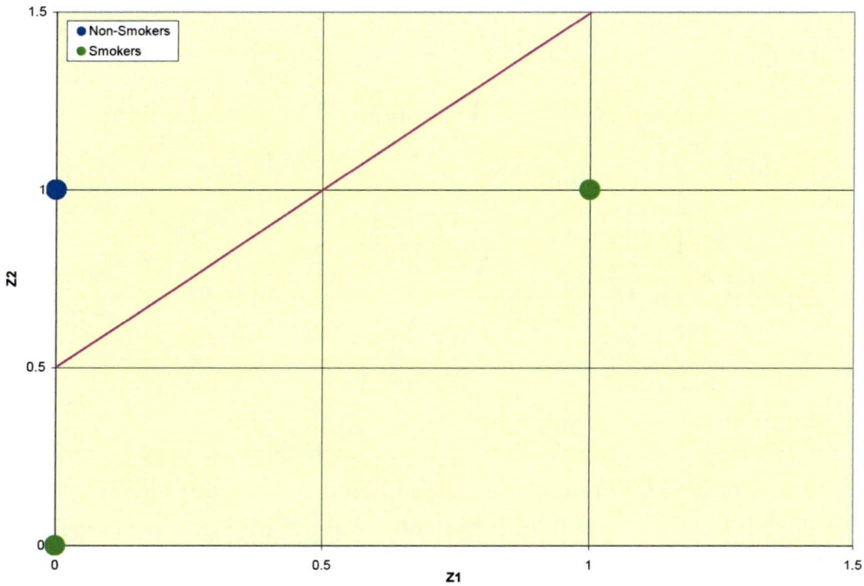

Line 4 (containing points $(0,0.5)$ and $(0.5,1)$):

$$w_{41} \cdot 0 + w_{42} \cdot 0.5 + w_{40} = 0 \wedge w_{41} \cdot 0.5 + w_{42} \cdot 1 + w_{20} = 0$$
$$\Rightarrow w_{42} \cdot 0.5 + w_{40} = 0 \wedge w_{41} \cdot 0.5 + w_{42} + w_{40} = 0$$
$$\Rightarrow w_{42} = -2w_{40} \wedge w_{41} = 2w_{40}$$

One of the possible triplets is $(w_{21}, w_{22}, w_{20}) = (-1,0,288)$.

Now the Z4 column has much more sense – after estimating the formula of the fourth line, smokers and nonsmokers are clearly separated. Although the collected data does not contain any input with Z values 1-0-0, the fourth line shows that potential data in this segment would be considered a smoker.

Finally, the last step is to separate the NSCLC cases from smokers. That is achieved by comparison of Z3 and Z4 coordinates – the unique evidence that the patient belongs to the NSCLC group is to have both Z3 and Z4 values equal to 1. This is equivalent to logical AND problem that is linearly separable so there is no need for any extra neurons in processing layer (one of the triplets for Line 5 coordinates are (1,1,–1.5)).

In final design there are two output neurons. The following model is implemented in VisSim modeling software:

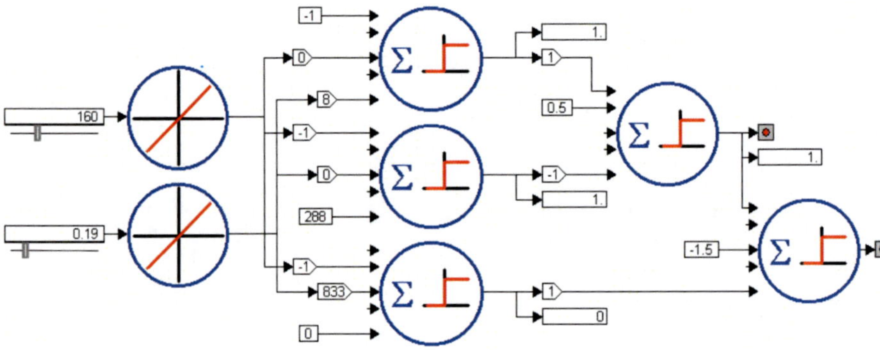

This is the complete network design, with two neurons in input layer, three processing neurons and two output neurons, one depending on the other. The apoptotic capacity can be inserted by the user through the scale from 0 to 400 and the apoptotic index is inserted by the scale from 0 to 1 (with two-digit accuracy). All the weights from the table are presented as coefficients that multiply the neural signals to the next neuron. At the end, there are two signal lights – the above one to state if the patient is a smoker or not, and the second one to state if it is the NSCLC case (blue light is negative and red light is positive value). For this certain preview it is the case of patient who is just in the control smoker group.

Dose Response and Hormesis-Threshold Model in Smoking and Lung Cancer under Exposure to Low-Dose Radiation

Another possibility is to observe a graph obtained by a neural network method as a dose-response relationship, regarding Z values which represent smoking exposure (Minic et al., 2011; Zunic, 2015; Zunic and Rakic, 2015).

Smoking exposure is a default character of the dose-response graph (Figure 15).

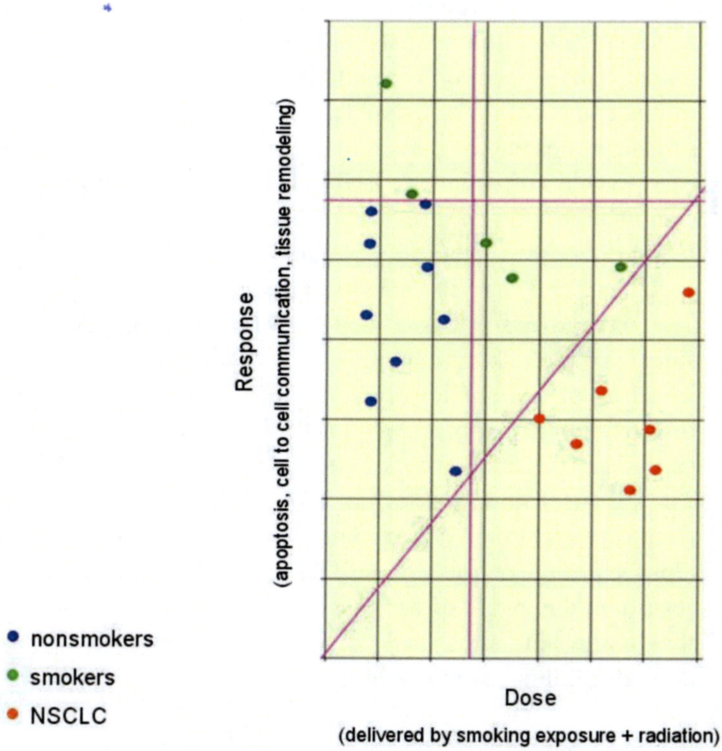

Figure 15. Dose-response relationship in smoking and low-dose-radiation exposure based on a neural network method. Dose (along x line) represents smoking and low-dose-radiation exposure. Response (along y line) represents the response of tissue and reflects the complex network of cell to cell interactions and tissue remodeling, including the balance between apoptosis and apoptotic clearance.

The graph in Figure 15 assumes two parameters: AI (along x line) and AC (along y line).

The advantages of the proposed neural network model for dose-response, based on the apoptotic parameters under smoking exposure and low-level-radiation doses, are:

1. The model is based on minimally invasive bronchoalveolar lavage procedure. Whereas histopathology analysis of tissue samples is spatially limited, bronchoalveolar lavage contains cells from the less restricted area of bronchoalveolar space, which is more representative model for screening, considering lung cancer multicentric and multistep process. The usefulness of this approach for the purposes of screening lies in the possibility to analyze the BAL samples at an early stage of clinical trial of patients, during bronchoscopy examination.
2. It is a low-cost method (cytological examination of TUNEL stained specimens and analysis by light microscopy).
3. The method represents a step forward in achieving individualized screening and risk estimation.
4. In terms of radiobiology, it contributes to a better orientation and distinction of protective and damaging mechanisms during carcinogenesis evoked by long-term inhalation of contaminated air.
5. The proposed model does not require an exact measuring of tissue doses in conditions of exposure to low doses of radiation, especially alpha emitting radioisotopes. This method enables quick orientation regarding the extent of damage of complex regulatory mechanisms in the tissue in situ, and indirectly indicates the existence of premalignant or malignant lesions.

The comparable effects of radionuclides from tobacco smoke, and radiofrequency electromagnetic field in the context of mitotic abnormalities, chromosome aberrations, micronuclei and mitotic index (Pesnya and Romanovsky, 2013), imply the opportunity to use the presented model for the estimation of cytotoxic and genotoxic properties of numerous radiogenotoxic agents.

Smoking exposure values are significantly different in smokers and nonsmokers, but they do not provide enough information regarding the lung cancer risk among smokers. The inclusion of some parameters that represent the apoptotic ability of tissue, by using the neural network model, makes it possible to describe the tissue regulation under the influence of tobacco smoke and low doses of radiation.

The advantages of neural networks are numerous, not only in terms of time. They also significantly reduce the calculation errors and boost learning and data processing skills.

The proposed model can be used for the purposes of screening and it is a step forward in achieving a personalized medicine approach in the early diagnosis of lung cancer.

Our findings correspond to the hormesis model of tissue response to radiation. This model deviates from the threshold model and assumes that radiation in higher doses increases the risk of cancer, but not linearly. In the best case, the model implies the existence of a threshold that cannot be clearly defined because of the tune changes in individual immunocompetence of tissue.

We conclude that low-dose radiation induces mixed hormesis and the threshold tissue response.

Chapter 7

METABOLOME CHANGES IN LOW-DOSE RADIATION EXPOSURE

According to a recent publication authored by Djorđevic and coworkers (2016), the properties of atmospheric aerosols show spatial and temporal variability. Dependent on the elemental composition, aerosols can induce cooling, or warming effect, variability of the scattering and absorption of light by airborne particles. There is a vast body of literature data, sometimes contradictory, relating to different sampling and different methods for measuring environmental contamination in different geographical locations (Popovic et al., 2008; Devi et al., 2012). Djorđevic and coworkers (2016) investigated the physical and chemical characteristics of continental urban aerosols whose samples were collected in the urban area of Belgrade. High values of aluminum, they reported, could be associated with the anthropogenic origin, most probably with global geo-engineering regarding the ongoing global climate modification programs.

Because of its permanent contact with the external milieu, the airway epithelium is frequently injured and itself repaired (Zunic et al., 2007). Regardless of the source of injury, lesions can vary from the loss of cell impermeability as a result of tight junctions opening, to a more or less shedding of the surface airway epithelial cells. After airway epithelial injury, the basement membrane may be completely denuded, or partial shedding of the airway surface epithelium can be observed with only clusters of basal cells remaining attached to the basement membrane. Immediately after injury, the airway epithelium initiates a repair process in order to restore its barrier integrity. In vivo models point to several common sequential processes of epithelial repair and regeneration as follows: 1) cell spreading and migration;

2) pre-mitosis and dedifferentiation followed by squamous metaplasia; 3) progressive redifferentiation (partial 2-3 cell layers); 4) full differentiation with ciliogenesis regeneration of ciliated cells (Puchelle and Zahm, 1996; Puchelle, 2003).

During the process of tissue repair, as well as tissue injury, extracellular matrix is included as an important counterpart due to its morphological and functional connections with cellular elements. Extracellular matrix is not an inert substance, but is turned over at a considerable rate every day, and there is a subtle equilibrium between synthesis and degradation. It is estimated that 10% of total pulmonary collagen is degraded and newly synthesized every day. Fibrosis is characterized by the abnormal accumulation of extracellular matrix in the interstitium resulting in impaired and excessive tissue repair. This process may be initiated by acute or chronic injury by a key profibrotic cytokine that promote myofibroblasts differentiation, enhance synthesis of collagen and other matrix components (Kolb et al., 2002).

Environmental noxious agents and host factors (genetics) in interaction and in relation with oxidative stress, abnormal inflammation, tissue damage, potentiate abnormal tissue repair affecting the tissue remodeling (Siafakis and Tzortzaki, 2002). Smoke impairs lung repair mechanism targeting enzymes which are dealing in the synthesis of components of extracellular matrix (Osman et al., 1985; Nakamura et al., 1995), and disrupts procedures that are able to restore tissue structure. This may lead to peribronchial fibrosis and narrowing, particularly at the site of small airways (Shoji et al., 1990).

The effects of chronic cigarette smoking on lung cell function are described as about 20 things different when compared with nonsmokers (Reynolds, 1987). Smoking may alter natural killer cell activity (Hughes et al., 1985). X-ray microprobe analysis identified particles related to smoking (kaolinite and mica) (Johnson et al., 1986). The accumulation of free fatty acids, phospholipids and triglycerides has also been discovered by cytochemical analysis in macrophages of smokers with developed arteriosclerosis (Plowman and Flemans, 1981). Inflammatory reactions in the lungs may be influenced by the local lipid environment (Hughes and Haslam, 1990).

When smoking is combined with other risk factors, the risk of developing lung cancer is increased, as we mentioned above (Zunic et al., 2007; Zunic, 2015).

Smoking is responsible for remodeling of the airways. Remodeling comprises at least three processes: adhesion, migration of cells and angiogenesis. After bio-trauma by (inhaled) a noxious agent from external milieu, pulmonary tissue reaction can result in changed micro-architecture with possible ways of resolving, oncogenesis or autoimmune process. Professional phagocytes (neutrophils, eosinophils, monocytes and tissue macrophages) are involved in the specific immune responses in the regulation of leukocyte transfer from the blood into inflamed tissue. As the pulmonary-tissue-resident cells, AMs contribute in the clearance of tissue during the process of programmed cell death of non-resident cells, recruited from the blood to lung during pathogenic process (Zunic et al., 2007).

Different stimuli, both physiological and pathological, can induce apoptosis in cells (Zunic et al., 2007). Apoptosis was observed in negatively selected T cells, irradiated lymphocytes, cytotoxic T-lymphocyte target cells, as well as in regressing tumors. Apoptosis is a morphological description seen in many, but not all examples of cell death. Not all programmed cell deaths occur by apoptosis, and some examples of apoptosis occur in the absence of new gene expression, such as that induced by cytotoxic T lymphocyte killing (Schwartz and Osborne, 1993; Jaattela and Tschopp, 2003). Within the tissues, most apoptotic bodies are rapidly phagocytized either by resident macrophages or neighboring cells and are degraded within phagosomes by lysosomal enzymes derived from the engulfing cell (Harmon et al., 1988).

Since then, we have defined a new model which could be of importance for cancer, as well as for preneoplastic conditions (Zunic, 2015, Zunic, 2016). Its role might be anticipated in the complex signaling-regulatory network, lying in damaged tissue micro-architecture in malignant and nonmalignant lung diseases (Zunic et al., 2007) (Figure 15).

Apoptosis is considered as a protective mechanism that limits lung injury (Oritz et al., 1998). It is an active form of programmed cell death that plays an essential role in development and survival by eliminating damaged or otherwise unwanted cells. Impaired regulation of apoptosis leads to a variety of pathological conditions, such as neurodegeneration, autoimmunity, chronic inflammation, AIDS and cancer (Yuan et al., 2002). Cigarette smoke interferes with neutrophile apoptosis, resulting in a delay, or prevention of programmed cell death and in an increased number of neutrophils undergoing primary or secondary necrosis (Schaberg, 2000). Lymphocyte apoptosis is essential for a proper functioning of the immune system. Among other functions, it is responsible for the homeostasis of immune cells and plays a key role in the elimination of auto-reactive lymphocytes. Studies characterizing the basic apoptosis signaling machinery have begun to reveal the molecular control of the processes modulating lymphocyte apoptosis (Scaffidi et al., 1999).

Cancer is a result of multiple gene-environment interactions occurring over several decades. During tumor development the cell accumulates multiple genetic changes, which generate the transforming genotype, i.e., a cell with increased genomic instability. Lung cancer is a useful model for the study of the interplay between genetic factors and environmental exposure since the primary etiology is well established. (Haugen et al., 2000).

Gene-environment interaction in cancer induction is more important than effect by a specific gene, environmental exposure, or gene-environmental interaction (Shields and Harris, 2000). During the smoking, whole tissue is exposed to carcinogenic insult and is at increased risk for multistep tumor development with preceding or accompanied premalignant lesion to tumor. Lungs of long term smokers (greater than 20 pack-year smoking history) show significant evidence of genomic instability, and this instability can be detected through the accessible bronchial tree, even when bronchial metaplasia is not evident (Markovic et al., 2008). Amount of chromosome polysomy reflects intensity and extent of tobacco exposure (Hittelman, 2001).

Our model, related to low-dose radiation and smoking exposure, was based on the group of smokers whose average smoking exposure (pack-years) values were <2, and patients with non-small cell lung cancer > 40 (Zunic et al., 2007). Bearing in mind our earlier results (Zunic et al., 2007), and the fact that uranium is inevitably present in the environment and inhaled air, originating from nuclear accidents or its military use, we could speculate about the metabolic effects of depleted uranium.

The effect of high-dose radiation is linked with the higher probability of DNA damage. At higher energies, a disruption of the double helix is more possible. Low radiation doses lead primarily to ionization events. Ionization is the most evident at the membrane level. The cell membrane is a lipid bilayer, and the endomembrane system divides into the multiple cellular compartments. The membrane's lipid molecules make an ideal structure that leads to the initiation and propagation of ionization. If the integrity of the membrane is distorted, the cell becomes vulnerable to the ionic and electrical gradients.

Along with other roles, the cell membrane is a means of communication of pores associated with the membrane of adjacent cells. Changing the nature of matter and the transfer of signals between cells, most likely induces the bystander effect. A large number of cells, which are located around the cell targets, is experiencing a transformation that leads to reparation, or decline, with a subsequent programmed cell death. On the other hand, the affected cells usually become lethally changed.

A disturbance of cell membrane or endomembrane system results in the damage of function and structure of cell compartments: it is possible that nuclear material, including DNA, enters the cytosol.

The disturbance at the level of mitochondrial membranes can damage the electron transport chain and oxidative phosphorylation ... On the other hand, an inadequate energy yield in the cell leads to the metabolic transformation of cells. A glycolytic phenotype becomes transformed into a lipogenic phenotype with the further transformation of cells that leads to the initiation of neoplastic lesions in the prolonged course (Menendez and Lupu, 2007).

There is some evidence concerning the changed subcellular organization, as well as disturbed ionic properties of cancer cells (Kroemer and Jaattela, 2005). Cancer cells show a redistribution of lysosomes from the nuclear periphery to the cytoplasmic membrane. The lysosomes are larger and show increased expression of cathepsins and increased iron (Fe^{2+}) content. Their pH is also reduced. Together with matrix metalloproteinases and the plasminogen-activator system, secreted cathepsins might participate in the degradation of the extracellular matrix, thereby enhancing cellular motility, invasion, angiogenesis and overall cancer growth. In addition, oxidative stress together with intra-lysosomal iron generates oxygen radicals. This might trigger lysosomal-membrane permeabilization through oxidation of membrane lipids, resulting in the destabilization of the lysosomal membrane.

Changes in the interactions of specific ions with biological matrices, lead to changes at the cellular level (Blanchard and Blackman, 1994). We have already discussed the ion resonance of metaloenzymes and other biomolecules containing a metal ion for interaction with magnetic field. Spectroscopic evidence suggested that UO_2^{2+} cation itself could be bonded to one D-glucose molecule and to two H_2O, resulting in six-coordination around the uranium ion (Tajmir-Riahi, 1988). This complex includes C-2 hydroxyl group of glucose, which is necessary for the further flux of glucose carbons through glycolytic pathway. As in case of F-18, a glucose molecule can be metabolized until the level of triose-phosphates. Any energy yield by glycolysis is consequently diminished (Zunic, 2015). A study by Koban et al., (2002) presents the results of uranium complexation with glucose-6-phosphate and fructose-6-phosphate. The Uranyl sugar phosphate species $UO_2(ROPO_3)$ (where R is either glucose or fructose) can be formed. Consequently, the lack of energy yield from both glucose and fructose ensues. Glucose-6-phosphate is one of the crucial molecules in metabolism. It is achievable that other metabolic pathways providing energy within the cell could be corrupted (Zunic, 2015). The catabolism of glucose and fructose, two important molecules of metabolic fuels, can be stopped by uranium. Consequently, as experimental studies showed, there would not be an energy yield.

We tried to estimate the metabolic, ionic and enzyme properties of alveolar macrophages, which represent in situ biochemical, immunological and signaling processes. It is in a doctoral thesis, entitled "Cytochemical parameters of alveolar macrophages in sarcoidosis and malignant lung diseases" (Belgrade University, Medical Faculty, 1993) (Zunic, 1993), that the cytochemical analysis of alveolar macrophages using semiquantitative indexing and scorring method was first described.

Originated from precursor stem cell of bone marrow, a monoblast becomes monocyte after entering the circulation. Upon entering the tissue, monocyte undergoes morphphological and functional changes giving a mature macrophage. After the activation, they function as antigen-presenting cells and promote lymphocyte activation. Macrophages are found in all organs and connective tissues, and are differentiated by pathogens into local-specific cells, like alveolar macrophages in lungs, osteoclasts in bone, microglia in the central nervous system (Fireman, 1996). Alveolar macrophages were observed as tissue resident cells which developed in pulmonary tissue in situ.

As the respiratory tract is the main path of internal contamination with DU particles from inhaled air, it is important to understand the role of AM for a few reasons: 1) due to their specific appearance and size, these cells are easily recognizable and convenient for cytochemical study; 2) AM cannot artificially appear in BAL due to artifacts during bronchoscopy procedure; 3) their immunological properties, metabolic and ionic peculiarities and enzyme activity represent local tissue background.

After BAL fluid separation, cytocentrifuge preparations have been made on glass slides, air dried and then stained by the standard method described by Hayhoe and Quaglino (1980) and modified by other authors (Cvetkovic, 1981; Radak, 1983; Djordjevic-Denic et al., 1987; Zunic, 1993) for the semiquantitative estimation of cytochemical reaction positivity of AM for:

- Nonspecific esterases (NSE):
  - Alpha naphthyl acetate esterase (ANAE) and
  - Butyrate esterase (BUT)
- Chloroacetate esterase (CHL)
- Acid phosphatase (ACP)
- Perls' reaction (for estimation of intracellular iron)
- Sudan black B reaction (SBB) (for lipids estimation)

- PAS reaction (for estimation of intracellular glycogen).

CYTOLOGICAL ANALYSIS OF THE BAL SAMPLES

Preparations were stained with May-Grunwald-Giemsa, mounted with DPX and covered with cover slipps. Differential cell counting was made from a count of 500 cells using light microscopy [x 40 (dry)] and [x 100 (immersion objectives)], in random fields and each cell type was expressed as a percentage of the total cells present regarding next cell types: neutrophils segmented, eosinophils, mast cells, macrophages, and lymphocytes.

CYTOCHEMICAL ANALYSIS OF AM

The semiquantitative cytochemical method includes the translation of an impression about the intensity of cell staining to numerical information, respecting intracellular distribution, size and intensity of the stained products of cytochemical reaction and in indexing and scoring method using Spearman's rank rule, with the theoretical range between 0-550. Rank of cytochemical reaction positivity of AMs has been estimated on 300 of AMs (Zunic, 1993) (Figure 16). F-test was used for statistical data analysis. The intensity of cytochemical reaction was evaluated semiquantitatively according to the following individual cell ratings:

- Positivity grade 0 of the cytochemical reaction relates to AM with absence of stained products of cytochemical reaction in the cell;
- Positivity grade I indicates AMs with a random distribution of diffuse small granules, like dust;
- Positivity grade II – granules are larger, single or in small groups, with cytoplasmic localization;
- Positivity grade III relates to AMs with large deposits of stained products in the cytoplasm and/or covering the nucleus; the same grade is given to AMs with diffuse staining of cell, but low intensity;
- Positivity grade IV relates to large deposits, which mask a major part of the cytoplasm, diffusely spread in the cells, which makes recognition of cellular structures impossible.

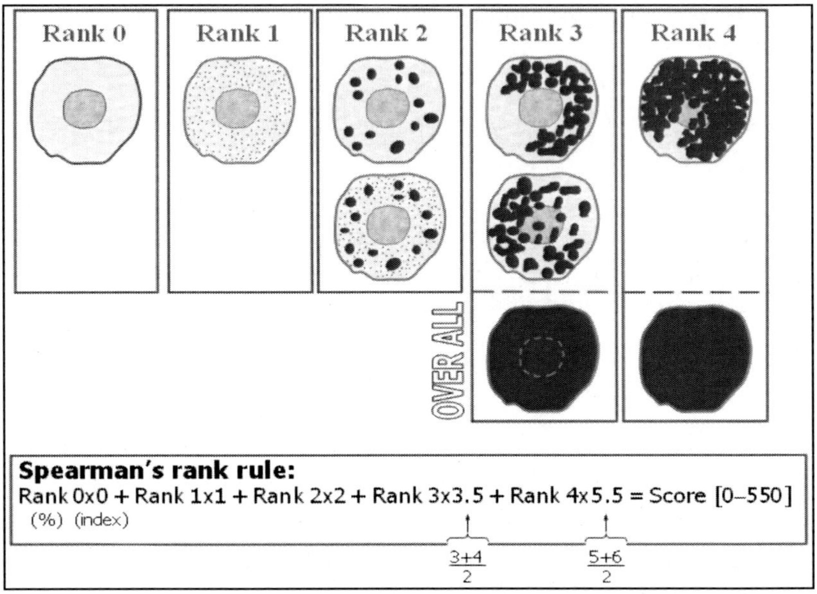

Figure 16. Schematic presentation of cytochemical analysis of AM. Figure shows schematic presentation of AM with different stages of cytochemical positivity: POSITIVITY GRADE 0 of cytochemical reaction relates to AM with absence of stained products of cytochemical reaction in the cell; POSITIVITY GRADE I indicates AM with random distribution diffuse small granules, like dust; POSITIVITY GRADE II - granules are larger, single, or in small groups, with cytoplasmic localization; POSITIVITY GRADE III contains AM with large deposits of products of cytochemical reaction, in cytoplasm and/or covering nucleus; the same rank has been given to the AM with diffuse staining of cell, but low intensity; POSITIVITY GRADE IV relates to large, high intensity deposits which mask major part of cytoplasm. The same rank has been given to the AM with diffuse staining of the cell, what makes unable recognition of cellular structures. Reproduced with permission of Zunic et al., (2007).

Per cent of AM with the same grade of cytochemical positivity was calculated, and then multiplied with appropriate index varying between 0-5.5; for example, per cent of AM with I grade of cytochemical reaction positivity was multiplied with 1, but per cent of AM without signs of cytochemical reaction positivity was multiplied with 0; for higher stages of cytochemical reaction positivity - stages III and IV, index was calculated respecting Spearman's rank rule and per cent of AM with the III grade was multiplied with 3.5, and AM with IV grade with 5.5. Then, the final score for a single slide calculated was with the theoretical range between 0 and 550.

The results of differential cell counting of bronchoalveolar lavage specimens and the results of semiquantitative cytochemical analysis of AM are presented in Table 11, and statistical significance of differences between investigated groups was shown in Table 12.

Table 11. Differential cell counting and cytochemical parameters of alveolar macrophages

		Nonsmokers Mean ± SD	Smokers Mean ± SD	NSCLC Mean ± SD
BAL cells (%)	Neutrophils	2.16±1.51	1.05±0.82	6.58±4.36
	Eosinophils	2.54±1.92	0.75±0.54	26.55±14.39
	Mast Cells	0.52±0.37	0.37±0.30	0.24±0.33
	Lymphocytes	19.50±4.58	15.90±3.76	25.92±9.08
	Alveolar Macrophages	65.86±5.77	75.18±6.25	35.21±23.79
AM cytochemical parameters	ANAE	424.78 ± 128.59	490.17 ± 81.10	482.00 ± 99.30
	BUT	182.85 ± 33.64	248.60 ± 174.57	214.44 ± 49.07
	CHL	387.44 ± 68.81	436.16 ± 93.39	459,00 ± 70.16
	ACP	377.00 ± 90.19	444.17 ± 69.04	365.00 ± 97.94
	Iron	237.56 ± 71.89	237.67 ± 80.24	353.72 ± 71.89
	PAS	514.56 ± 36.79	513.34 ± 32.41	493.00 ± 61.19
	SBB	366.56 ± 139.07	456.17 ± 131.90	418.28 ± 98.21

Adapted from primary data of Zunic, (1993).

Neutrophils were significantly increased in NSCLC (6.58±4.36) in comparison with nonsmokers (2.16±1.51) and smokers (1.05±0.82) ($p < 0.05$). There was no significant difference between nonsmokers and smokers for neutrophils as well as for lymphocytes and mast cells, between investigated groups. Alveolar macrophages were significantly higher in smokers in comparison with nonsmokers (75.18±6.25 vs. 65.86±5.77, $p < 0.05$), but decreased in NSCLC patients in comparison with both control groups (35.21 ± 23.79, $p < 0.01$). BAL eosinophils of NSCLC patients were quite significantly increased in comparison with control groups of nonsmokers and smokers (26.55±14.39, vs. 2.54±1.92 for nonsmokers, $p<0.001$, vs. 0.75±0.54 for smokers, $p<0.01$ respectively). There was a significant decrease of eosinophils in BAL of smokers in comparison with nonsmokers ($p<0.05$). BAL lymphocytes were significantly increased in patients with NSCLC in comparison with smokers (25.92±9.08 vs. 15.90±3.76, $p<0.05$). Although in smokers lymphocyte count is lower than in nonsmokers, no significant difference was found. The decrease of lymphocytes is caused in part by concomitant increase of BAL macrophages, in control smokers in comparison with nonsmokers.

Table 12. Statistical significances of differences for BAL cells percentages between investigated groups

Groups compared	Nonsmokers vs Smokers	Nonsmokers vs NSCLC	Smokers vs NSCLC
Neutrophils	ns	$p < 0.05$	$p < 0.05$
Eosinophils	$p < 0.05$	$p < 0.001$	$p < 0.01$
Mast Cells	ns	ns	ns
Lymphocytes	ns	ns	$p < 0.05$
Alveolar Macrophages	$p < 0.05$	$p < 0.01$	$p < 0.01$

ns = non significant.

Adapted from primary data of Zunic, (1993).

Evaluation of results of semiquantitative cytochemical analysis of AM showed the significant increases of CHL and iron content in AM in patients with NSCLC in comparison with nonsmokers, both at level $p<0.05$ (Table 13).

Significant correlation between ACP and CHL was found in smokers ($r=0.99$, $p<0.05$), that is not present in nonsmokers. Human ACP in alveolar macrophages is tartrate resistant acid phosphatase, belonging to a group of widely-distributed and structurally highly-conserved group of iron-containing proteins (Moss, 1992).

Table 13. Statistical significances of differences for AM cytochemical parameters in investigated groups

Groups compared	Nonsmokers vs. Smokers	Nonsmokers vs. NSCLC	Smokers vs. NSCLC
ANAE	ns	ns	ns
BUT	ns	ns	ns
CHL	ns	$p<0.05$	ns
ACP	ns	ns	ns
Iron	ns	$p<0.05$	ns
PAS	ns	ns	ns
SBB	ns	ns	ns

ns = non significant.

Adapted from primary data of Zunic, (1993).

Tartrate resistant acid phosphatase (EC 3.1.3.2) is designed as type 5 acid phosphatases (Vincent and Averill, 1990). Tartrate resistant acid phosphatase is a member of the ubiquitously expressed enzyme family of the acid phosphatases. The physiological substrates for this enzyme have not been identified yet, but functional importance was based in the involvement in bone resorption and iron homeostasis (transport, metabolism) (Lamp and Drexler, 2000).

A possible functional role of tartrate resistant acid phosphatase gene product may be in the storage or the transport of iron (Fleckenstein, 1996). Intracellular iron content is involved in the regulation of the enzyme (Lamp and Drexler, 2000). The binuclear iron center that is present at the active site of tartrate-resistant ACP conveys catalytic activity to the enzyme molecule (Hayman and Cox, 1994).

Why might ACP be important for airway remodeling? ACP has a constitutive and regulatory role in the embryogenesis, and during the whole life this enzyme modulates remodeling processes in several tissues at the transcriptional level. The regulation of gene expression for tartrate resistant acid phosphatase depends on iron (Fleckenstein et al., 1996), cytokines (Bevilacqua et al., 1991) and hormones – estradiol at the mRNA level (Zheng et al., 1995). This regulation is precise: endocrine and in situ (fast, multifactorial). Tartrate resistant acid phosphatases are expressed in bone-resorbing osteoclasts (5b form), may be circulating marker of bone resorption, alveolar macrophages (5a form) or marker of inflammation (Halleen, 2006).

It has been proposed that during tartrate resistant acid phosphatase catalysed hydrolysis of phosphor-monoesters the iron center enables the effective binding of a phosphate substrate in the active site in an acid (pH 4.5-6) environment. However, evidence is accumulating for a second function for the binuclear iron center of tartrate resistant acid phosphatase, which is generation of ROS. Tartrate resistant acid phosphatase could catalyse the formation of free radicals (Halleen et al., 1998). Purple acid phosphatases (PAPs) are acid metallohydrolases that contain a binuclear $Fe^{3+}M^{2+}$ centre in their active site, where M is Fe or Zn (Antanaitis and Aisen, 1983; Averil et al., 1987; Doi, Antanaitis and Aisen, 1988). In mammals, these enzymes are also referred to as tartrate-resistant acid phosphatases The PAPs usually exhibit a pH optimum for the hydrolysis of phosphomonoesters in the range 5.5–6.0. In contrast to the Fe-Fe enzyme, the mixed metal derivatives are not rapidly inactivated by H_2O_2 (Beck et al., 1984). It seems that if metal ion bound to the active centre yield to higher activity of enzyme by increased iron bounding lead to further inactivation of enzyme (Zunic et al., 2007).

Some tumors from tissues where the activity of tartrate resistant acid phosphatase was recognized have the highest affinity for bone metastases. Osteoclasts with high TRAP activity are crucial for remodeling of bone tissue. AcP5b in osteoclasts and AcP5a isoenzyme subtype in AMs regulate adhesion and migration of cells with monocytic origin (Halleen et al., 1999). TRAP gene is a target of microphthalmia transcription factor. The microphthalmia transcription factor gene implies a critical role in regulating gene expression during osteoclast ontogeny (Luchin et al., 2000).

Although tartrate resistant acid phosphatase is recognized as a histochemical and biochemical marker of osteoclasts, and alkaline phosphatase in the osteoblasts (Burstone, 1959), there is evidence that bone forming cells, osteoblasts, and osteocytes also express a type 5 ACP (Lau and Baylink, 2003). Since 25% of the systemically administered uranium deposits in the skeleton are linked to the newly formed bone, the toxic effect of uranium would be causing an alteration of the differentiation process of osteoblasts and/or their precursors (Tasat et al., 2012).

Different patterns of cell-to-cell communications have been revealed in diverse groups of examinees characterized by a particular pathological condition. The investigated parameters refer to the integral parts of multi-factorial regulatory influences, arising from the metabolic, ionic and immunological properties of cells from bronchoalveolar space (Zunic et al., 2007).

The non-specific esterases of the human alveolar macrophages were localized at the outer side of the plasma membrane (Jaubert et al., 1978). Consequently, these esterases are ectoenzymes which may function as mediators of cell response to injurious agents from the outside. Monahan, Dvorak and Dvorak, (1981) reported that ANAE activity was present on the cell surface of lymphocytes, monocytes, macrophages, neutrophils, eosinophils, basophils, megakaryocytes, platelets, and blasts. Mononuclear phagocytes had multiple discrete foci of similar appearing ANAE-positive cytoplasmic-vesicle clusters that sometimes became confluent. ANAE activity was also found in the Gall bodies of human lymphocytes and in coated vesicles of macrophages. Cytoplasmic ANAE activity was increased in oil-induced guinea pig peritoneal macrophages. Both surface and cytoplasmic esterase activities had a neutral pH optimum. An identical distribution of reaction product was observed when alpha-naphthyl butyrate was employed as substrate.

BUT is important in short chain fatty acids metabolism as the principal energy source for epithelial cells growth arrest and differentiation pathways (Heerdt, Houston and Augenlicht, 1994). Sodium butyrate, a short-chain fatty acid naturally present in the human colon, is able to induce cell cycle arrest, differentiation and apoptosis in various cancer cells. This effect is antagonized by leptin (Rouet-Benzineb et al, 2004). Based on the findings published by Kim and coworkers (2004), that butyrate may be an effective sensitizer of (TNFα)-related apoptosis-inducing ligand, butyrate esterase may be estimated as with antiapoptotic effect. Kim and coauthors (2004) implicated the synergistic apoptotic effects of sodium butyrate and proteasome inhibitor MG132. Nevertheless, butyrate can be adaptor for mechanism through which environmental signals affect postnatal maturation of sympathoadrenal transmitter systems (Nankova et al, 2003).

The protein activities are compromised by the LET increase (Tasat et al., 2012). It has been clarified that alpha-radiation causes great loss of the enzyme activity (Available at: ASRC website). Consequently, key metabolic pathways can be targeted, as well as the cell oxidative metabolism.

As cells are more distant from the blood vessel, the oxygen diffusion limit increases (Menendez and Lupu, 2007). Random and irregular tissue damage and repair changes oxygen diffusion limit and proliferating cells change their metabolic phenotype from glycolytic to lipogenic. About 25 enzymes are involved in the metabolism of glucose to fatty acids (Menendez and Lupu, 2007).

In line with our findings (starting from Zunic, 1993, up to now), we proposed a model of biological effects of DU particles that can be found in the circulation, or in tissues, obtaining higher penetration power if they are micro- or nano-sized (Figure 17).

Busby (2013) argues that ionization is uniform across tissue. Under these circumstances, it is only a matter of probability whether a cell is intercepted by a track or not. For internal exposures (as we presented in Figure 17), the probability of interception of the track is clearly a function of the distance of the nuclide from DNA. In addition, internal exposures may be exposures to α tracks that carry significantly more ionization density. The range of most α tracks (which carry about of 5 MeV energy) is about 4 cell diameters (see Figure 17) and so, theoretically, the track dose to the cell from one decay is in the region of 500 mSv (Busby, 2013).

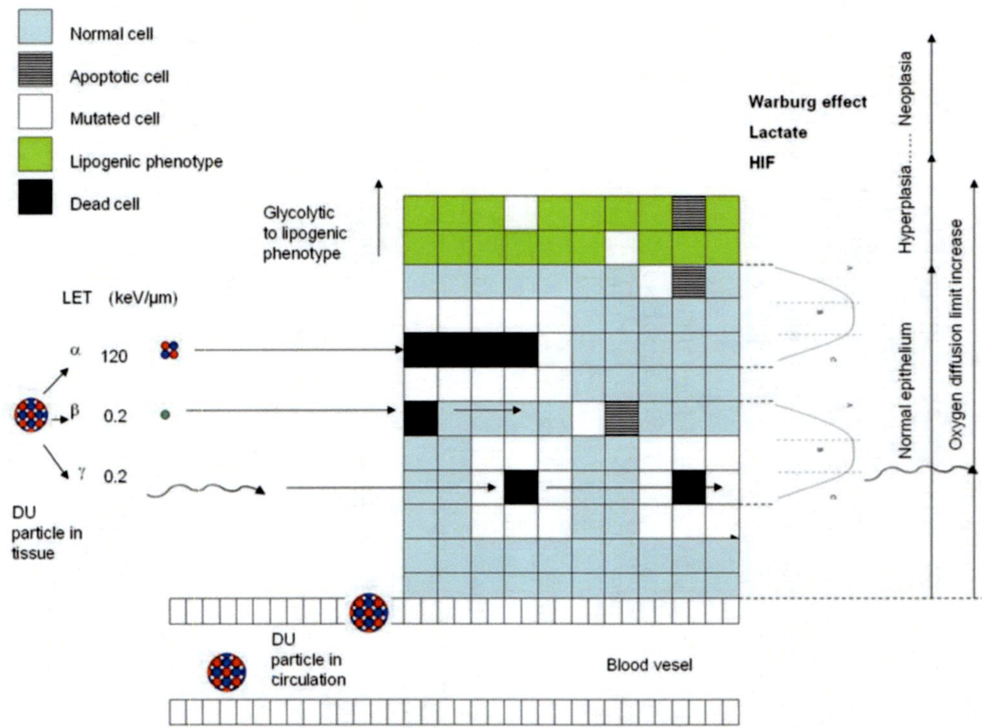

Figure 17. Schematic presentation of the interaction of α, β, and γ - radiation emitted by DU particle with tissue: A range of local energies (local dose) has little effect (A), a genetic effect (B) or a killing effect (C) on the tissue exposed to embedded DU particle. According to: Busby, C., 2013. Random and irregular tissue damage and repair changes oxygen diffusion limit and proliferating cells change their metabolic phenotype from glycolytic to lipogenic. According to: Menendez and Lupu (2007).

This is in the case of uranium. Linear energy transfer is one of the parameters to describe the ionization density, defined as energy deposited from radiation to materials per uint length or the radiation track (keV/μm). Higher LET radiation caused much severe DNA damage. Typical LET values are for γ-rays and β-particles, 0.02 keV/μm and for α-particle, 120 keV/μm (Available at: http://asrc.jaea.go.jp/ soshiki/gr/eng/mysite6/sub2.html). Uranium and DU are alpha emitters with high chemical affinity for DNA (Busby and Morgan, 2006).

This study cited the model that was established by Hohenemser and coworkers in 1986 in relation to the Chernobyl accident. The problem with the hot particle issue is that there will be a range of local energies (local dose) which will have either little effect (A), a genetic effect (B) or a killing effect (C) (Busby, 2013) (schematically presented at the tissue level in Figures 17 and 18). This may be the key point for understanding why low-radiation doses may induce a disproportionate cytotoxic effect. By observing (A), (B) and (C) fields that correspond to DU emitted corpuscular and electromagnetic radiation, it is apparent that random striking of high energy particle, or γ wave, with cells induce superposing effects if happened in lose vicinity. This means that cells which would be normal (A) or mutated (B) may be influenced by neighboring targeted or mutated cells. Due to superposition of (A), (B) and (C) fields around close targeted cells, much more lesions arise in surrounding tissue per volume, than in case of high doses of radiation.

DU emitted alpha particles induce enormous ionization, spending their huge energy along the short track in tissues (Figure 17). As a result, death of cells can be achieved. If pulmonary tissue is considered, alveolar macrophages are main phagocytes of tissue debris and remnants of dead or damaged cells. Tissue may repair this defect with normal or mutated cell. Due to the bystander effect, normal cells can be transformed into mutated ones. Some of these cells undergo apoptosis.

Smoking and lung cancer are associated with the increase of apoptotic rate; apoptotic clearance by AMs was significantly decreased in patients with NSCLC in comparison with nonsmokers and smokers ($p < 0.05$) (Table 9). After cell death, the remnants can be removed by AMs phagocytosis.

We clearly illustrated an apoptotic clearance by alveolar macrophages' phagocytosis. What about alveolar macrophages?

There was a significant decrease of percentage of AMs in NSCLC in comparison with nonsmokers ($p < 0.01$) and control smokers ($p < 0.01$) (Table 11 and Table 12).

We also documented a remarkable increase of iron in AMs of patients with NSCLC in comparison with control groups of nonsmokers (Table 11).

Accordingly, we postulate that ferroptosis could be a mechanism for iron-dependent form of non-apoptotic cell death of AMs in lung cancer (Figure 18).

Ferroptosis is an iron-dependent form of non-apoptotic cell death. Cellular energy depletion is not observed in ferroptosis.

AMs possess iron centers in their enzymes. During the prolonged tissue damage and inflammation, AMs gather iron, which can itself contribute to the process of cell damage, taking the role of molecular antennas. Then an electromagnetic field can interfere with biofrequencies leading to further damage of cellular integrity.

The Petkau effect was potentiated.

Other white cells are recruited from the circulation to remove damaged cells and tissue debris. Due to enormous phagocytosis, some of these cells acquire the appearance of the Lupus Erythematosus cells. Inflammatory reaction and mediators of inflammation, together with ROS, trigger *circulus viciosus*, resulting in the neoplastic transformation of tissue, based on mutated cells and time-dependent metabolic and ionic changes in pulmonary tissue (according to Figures 17 and 18).

The recognizing and quantification of biological changes at cellular or molecular levels in the living media may be more appropriate procedure for evaluation of low-dose radiation, than measurement by dosimetry techniques (Gatti and Montanari. 2004; Zunic, 2013; Zunic, 2013-1; Zunic and Rakic, 2015; Azimian et al., 2015). One of the reasons is that physical dosimeters are not capable to assess unplanned or accidental exposures from occupational, therapeutic or environmental sources (Azimian et al., 2015); the second one is the lack of appropriate method for exact measuring of internal doses (Gatti and Montanari. 2004). In the study with low doses of ionizing radiation (0.05 Gy and 0.5 Gy), the bystander cells demonstrate initiation of the apoptotic cascade by the up-regulation of p53, Bax, Bcl-2, initiator caspase 2 and effector caspase 6. Due to down-regulation of the effector caspases 3 and 7 in their gene expression levels at 0.05 Gy and 0.5 Gy, cell death may not be executed to the final stages in spite of the up-regulation of pro-apoptotic and initiator genes (Furlong, 2013).

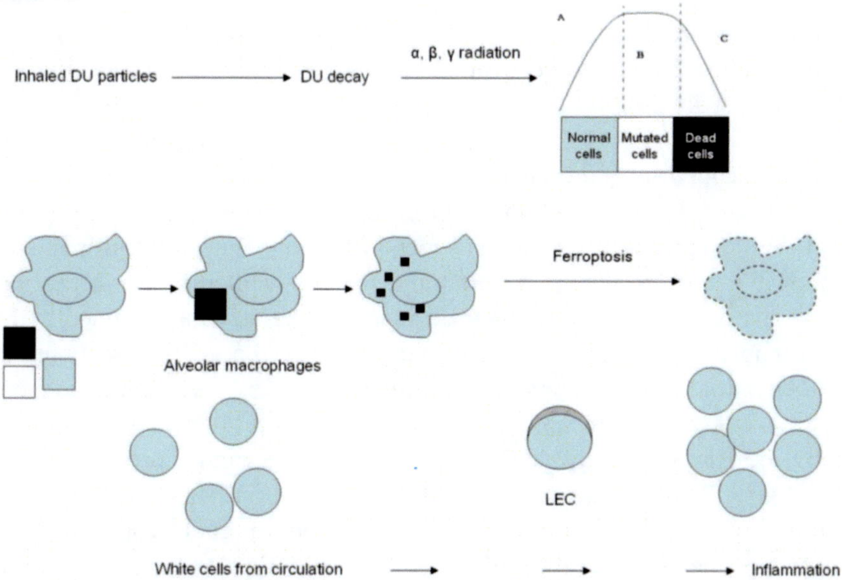

Figure 18. DU induced tissue response. A range of local energies (local dose) has little effect (A), a genetic effect (B) or a killing effect (C) on the tissue exposed to embedded DU particle. After cell death, the remnants can be removed by alveolar macrophages' phagocytosis. During the prolonged tissue damage and inflammation, alveolar macrophages phagocytize tissue debris, gather iron and undergo ferroptosis. Other white cells are recruited from the circulation to remove damaged cells and tissue debris. Some of these cells acquire the appearance of the Lupus Erythematosus cells (LEC). According to Busby, C., 2013. and Zunic, S., (2013-1).

The members of the Bcl-2 gene family are included in regulation of the cell response to radiation and regulation of apoptosis. Ionizing-radiation–induced apoptosis involves the Bcl-2 gene family and is mediated by extrinsic and intrinsic pathways, whereas the latter is the principal cause of apoptosis (Azimian et al., 2015). The Bcl-2 family contains pro-apoptotic members Bid, Bax, and Bak, which trigger protein release from mitochondria, and antiapoptotic members, Bcl-2 and Bcl-xL, which inhibit protein release (Kuwana et al., 2002, Furlong, 2013). Bax and Bak are crucial executioners of the intrinsic pathway of apoptosis. Miyashita and Reed (1995) concluded that the p53 gene was an inducer of apoptosis by transactivating expression of the Bax gene.

A study by Azimian et al., (2015) evaluated the dose-dependent and time-dependent patterns for Bcl-2 and BAX gene expression. Low doses of gamma radiation can induce early (after 4 h) down-regulation of the BAX proapoptotic gene in freshly isolated human peripheral blood mononucleated cells. These changes were restored to near normal levels after 168 hours (Azimian et al., 2015).

The intrinsic pathway of apoptosis by adjusting the release of mitochondrial proteins controls the occurrence of apoptosis (Azimian et al., 2015). Recent data of Große and coworkers (2016) showed that during apoptosis, Bax translocates to the mitochondria and mediates the permeabilization of the outer membrane, thereby facilitating the release of pro-apoptotic proteins. The authors used super-resolution data, obtained by STED nanoscopy, to provide direct evidence that large Bax-delineated pores in the mitochondrial outer membrane are crucial for Bax-mediated mitochondrial outer membrane permeability in cells. This process includes clustering of activated Bax molecules, and also assembling with other proteins, including Bak, into ring-like structures in the mitochondrial outer membrane. The authors explained that both, Bax and Bak, are essential for mitochondria-mediated apoptosis. In healthy cells, Bax and Bak are constantly shuttled between the mitochondria and the cytosol. Because of different rates for the retrotranslocation from the mitochondria to the cytosol, Bak resides predominantly in the mitochondrial outer membrane, whereas the majority of the Bax molecules are in the cytosol. The Bax activation blocks shuttling into the cytosol during apoptosis, and consequently Bax accumulates on the mitochondria, and in a form of oligomers inserts into the outer membrane. Large "Bax clusters" on the mitochondria develop at later stages of apoptosis. Bax induced outer mitochondrial membrane pores enable release of cytochrome c as well as larger molecules. There is a possibility that activation of Bax and Bak may result in membrane bending and the formation of lipidic pores, allowing protein instead of forming defined channels or pores allowing protein (Große and coworkers, 2016).

Results of Kuwana et al, (2002) showed that translocation of large mitochondrial proteins during apoptosis require cardiolipin. Supramolecular openings in the outer mitochondrial membrane are promoted by BH3/Bax/lipid interaction. Spatial organization of Bax in apoptotic cells using dual-color single-molecule localization-based super-resolution microscopy showed that active Bax clustered into a broad distribution of distinct architectures, including full rings, as well as linear and arc-shaped oligomeric assemblies that localized in discrete foci along mitochondria. These rings and arcs assemblies of Bax perforated the membrane, as revealed by atomic force microscopy in lipid bilayers.

All these mechanisms may be summarized as follows: although high doses of gamma radiation can cause apoptosis, low doses of gamma radiation cause reduction of lymphocyte radio-sensitivity which relates to the increase of the Bcl-2/BAX ratio. According to Azimian et al., (2015), the up-regulated expression of Bcl-2 is the proposed mechanism for the radioresistance effects of low dose exposure.

Ionizing radiation induces an apoptotic process which involves disorganization of outer mitochondrial membrane, resulting in its increased permeability for some of protein molecules, constituents of electron transport chain. Consequently, the integrity of transmission of electrons through the respiratory chain was damaged, as well as proton and electrochemical gradients, for which maintenance the integrity of mitochondrial membranes and inter-membrane space were required. Damage at the level of electron transport chain, or oxidative phosphorylation, is resulting in energy depletion, or less ATP was produced.

Together with mitochondria, other cell organelles, including the endoplasmic reticulum, lysosomes, and the Golgi are integrated into pro-apoptotic signaling (Rajiv, 2014).

Deregulated redox signaling, or mitochondrial dysfunction induced by ionizing radiation, may induce autophagic cellular response. Autophagy process relates to the degradation of cytoplasmic components within lysosomes (Mizushima, 2007). Autophagy can be related to hormetic tissue response to low doses of stressors, including ionizing radiation (Szumiel, 2012). We showed the existence of *in vivo* hormetic response to low-dose radiation, which was estimated by apoptotic parameters (Figure 15) (Zunic et al., 2007; Zunic, 2015). We mentioned above that alveolar macrophages' metabolic, ionic and enzyme properties were changed, as well as apoptotic markers. In our studies, apoptosis was considered together with apoptotic clearance by alveolar macrophages' phagocytosis (Zunic et al., 2007; Zunic, 2015). As inhaled air is the main source of internal contamination, further research on this topic is valuable, especially in terms of overcoming inter-individual variability. We proposed a simple model based on apoptotic parameters and artificial network method for individualized estimation of tissue response to low-dose & tobacco exposure (Figure 15).

A balance between the signaling functions and damaging effects of ROS seems to be the most important factor that decides the fate of the mammalian cell after exposure to ionizing radiation (Szumiel, 2012). Apoptosis and autophagy are two types of programmed cell death.

Apoptotic cell death, which is a non-inflammatory cell response, and necrotic cell death, which may be inflammatory, are two extremes, with overlap between them. When a cell dies by a typical apoptotic process, usually there occurs a late-phase necrosis. The plasma membrane integrity is not maintained in necrosis; the necrotic cell disintegrates and releases its cellular content, including lysosomal enzymes, into the extracellular space, which is a trigger for an inflammatory response. Although lower intracellular ATP level favors necrosis over apoptosis, as apoptosis needs energy, regulated necrosis can occur with high intracellular ATP levels (Rajiv, 2014).

According to Rajiv, (2014), ionizing radiation can induce cell necrosis directly, by damaging influences to the cell. Primary necrosis is triggered through unregulated processes of membrane and cytosolic destruction under extreme conditions (inadequate secretions of cytokines, nitric oxide and reactive oxygen species and calcium cytotoxicity). Secondary necrosis occurs in late apoptotic cells which fail to be engulfed by macrophages. Cells are not phagocytosed and thus lose membrane integrity, cease to be metabolically active and release their cytoplasmic content out into the extracellular matrix.

Cell necrosis is based on multi-organelle dysfunction related to mitochondrial dysfunction (which results in ATP depletion) and dysfunction of ER and calcium homeostasis (which resulted in increased ROS level, further damage of membranes, profound ATP deprivation, finally leading to necrosis). The type and intensity of noxious signals, ATP concentration, cell type, and other factors determine how cell death occurs (Rajiv, 2014).

Finally, there is a wide spectrum of cellular events which can be evoked by radiation. According to our results referring to in vivo cell processes, we look into at least three ways cells can die. Our model assumes interaction of alveolar macrophages with other white cells in bronchoalveolar lavage specimens. We presented immunological and metabolic competence of alveolar macrophages that encounter and fight with incoming noxious agent. Apoptosis, ferroptosis and necrosis (with coexistent inflammatory response in the tissue) are main destinies of a cell which is exposed to environmental noxious agents, including external and internal sources of ionizing radiation. The response of the cell or tissue to low-dose radiation depends on dose and duration of exposure. Not all cells in tissue with deposed DU particles are exposed equally. The processes of apoptosis, necrosis and coexistent inflammation may overlap and their demarcation could not be possible (Figure 17, Figure 18).

We discussed the results of pulmonary response to low-dose radiation induced by DU. Having in mind high internal penetrability of uranium micro-

or nano-particles, we achieve similar response in other tissues. Apoptosis is a process included in tissue remodeling. Consecutive necrosis/apoptosis/inflammation processes may be followed by proliferation/metabolic changes/metaplasia/neoplasia, according to Menendez and Lupu (2007) (Figure 17).

Multisystem and multisymptom occurrence referring to Gulf War syndrome or Balkan syndrome may be a result of the excess of cell death, or enhanced apoptosis which is a pathogenic mechanism in many diseases: cardiovascular (ischemia, myocardial infarction, heart failure, stroke), respiratory (asthma, COPD, infection, ARDS, interstitial fibrosis), neurodegenerative (Alzheimer disease, Parkinson disease), endocrine (diabetes mellitus, type 1 and 2), infective (HIV) and sepsis. Diminished apoptosis is one of key pathogenic mechanisms in the development of cancer, autoimmunity, persistent infections (Rayiv, 2014).

Chapter 8

ARE THE HEALTH EFFECTS OF DEPLETED URANIUM RECOGNIZABLE AND PREVENTABLE?

The explanation related to the limited effects of α-emitting nuclear weapons, including DU, was based to some extent on the fact that alpha particles have a short track in air. This paradigm has changed with the realization that nano- and micro-sized particles of DU could have a global atmospheric movement (Zunic and Rakic, 2015). The idea about spreading of uranium particles through air masses across the globe, arose from the results of measurement of air pollution (Busby and Morgan, 2006). The authors revealed a statistically significant increase in uranium in all the filters observed in the UK, beginning at the start of the Second Gulf War and ending when it ended.

Due to a global spreading of contaminated air masses, the inhalatory path of internal contamination is the most achievable. The lag time in understanding of biological effects, their extensiveness and health effects is a consequence of demanding procedure for exact detection of DU in the tissue (Bleise, Danesi, and Burkart, 2003). Later on, Gatti and Montanari (2004) concluded that the presence of DU particle in the tissue was not obligatory to determine whether a person was exposed. The lung can serve as a location of interactions between alpha particles and the lung immunocytes, where autoreactive T cells become reactivated and gain the competence to enter the CNS. The lung could therefore contribute to the activation of potentially autoaggressive T-cells and their transition to a migratory mode, as a prerequisite for entering their target tissues and inducing autoimmune disease (Odoardi et al., 2012). This fact makes a step forward towards understanding the multisystem involvement and multipathologies expression in case of DU contamination.

According to Priest (2001), in man, for chronic irradiation from an internally deposited radionuclide, the latency periods would typically lie in the range of 10 years to several decades. He concluded that "in view of this latency, tumors in individuals exposed for shorter periods – e.g., in servicemen exposed to depleted uranium in the former Yugoslavia within the past decade, cannot be attributed to radiation from depleted uranium". In this case, lag time was about 5 years (considering the bombing of Bosnia and Herzegovina in 1994/5, and later, the bombing of Serbia and Montenegro, 1999). Nevertheless, if we take into consideration a possibility that military personnel, as well as civilian population, in the former Yugoslavia have been exposed to air pollution containing DU from the Persian Gulf since 1990, then possibility of cancer induced by DU may be a real elucidation. Due to long pulmonary retention of 1,470 days, which is expected in the case of inhalation of uranium oxides, as Durakovic reported in 1999, a wide range of clinical manifestations can occur, depending on the individual predispositions of the exposed persons (Zunic, 2013-1, Zunic and Rakic, 2015). Experimental exposure to DU led to impaired coordination and movement performance in rats with multisystem damage including the brain (Seidemann et al., 2011). Zunic, (2013-1) discussed the possible pathogenesis of a multisystem and time dependent expression of signs and symptoms describing Gulf and Balkan syndrome. Gulf/Balkan War Syndrome was explained as a multicausal disease with multisystem involvement and time-dependent expression of symptoms from no cancerous diseases, to cancers in later phases, affecting soldiers, as well as overall civilian population (for more details on this hypothesis see conceptual map, Figure 10) (Zunic, 2013-1). Gulf or Balkan syndrome is the term describing the same medical entity. Repeated exposure to low doses of alpha radiation originating from the decay of internally deposed DU particles was understood as a main contributing factor to the onset of Gulf/Balkan syndrome.

All tissues with oxidative metabolism were targeted, particularly kidney and bone. DU causes electrical changes in the hippocampus of the brain, the area of memory and learning. Neurological/psychiatric manifestations of the disease are in part the consequence of demyelination, and together with other diseases with multisystem involvement are not rare. Psychiatric disorders are more frequent in the Gulf War veterans than in any other population, but without precise cause and effect relationship (Li et al, 2011).

During IPPNW conference in Belgrade (2015), studies by Kaatsch et al., (2008) related to leukemia incidence in children in the nuclear plant vicinity, and a study by Mathews and coworkers (2013) on medical exposure to radiation during CT, were emphasized as the examples of the health effects of chronic ionizing low-level radiation in civil circumstances. The overall risk is about 0.125 cancers per Sievert [of exposure] for medical purposes. Hand (2013) emphasized the importance of exposure to low-dose (medical) radiation during adolescence or childhood, which contributes to the increased risk of later cancer development.

The LEC phenomenon in BAL was understood as a radioprotective/ radioadaptive tissue response (Zunic, 2013-1). The tissue opposes long-lasting inflammation as well as diminished apoptotic clearance by AM by different adaptive mechanisms including the LEC phenotype expression. The finding of LEC+ pneumonitis in these two newborns we understand as possible evidence of early response of pulmonary tissue to DU particles' contamination which was transferred across the placenta from mother's circulation into a fetus. Interpretation of the presence of LEC in native BAL specimens changes one of the longstanding medical paradigms that LEC is mainly an in vitro phenomenon, which can be observed in patients with the diagnosis of autoimmune diseases (Zunic et al., 1996; Zunic, 2013; Zunic, 2013-1).

According to Sousa's (2016) report, prenatal stress, or excess exogenous glucocorticoid exposure, have been consistently linked to adverse health outcomes including low birth weight, neuroendocrine dysfunction and increased risk of infectious, cardiometabolic and psychiatric diseases in later life. An uncommon and little-studied type of cell in the lungs has been found to act like a sensor, linking the pulmonary and central nervous systems to regulate immune response in reaction to environmental cues (Branchfield et al., 2016). These two very recent publications shed new light on mass-catastrophic accidents, climate and weather extremes, wars... Not only different neuronal pathways are included in the transmission of heating and cooling signals to higher brain regions (Florence and Reiser, 2015)! Stressful stimuli in healthy subjects give rise to a consistent and reproducible activation of a set of brain regions (Sousa, 2016). "Neurosensorial-matrix" implies that the nature of the stressor determines the sensorial pathway that is initially recruited. According to Sousa's (2016) paper, if the stressor is of physical nature (for example, a painful stimuli, hypovolemia, exposure to inflammatory cytokines, hypoglycemia), activation of the brain stem nuclei, or of circumventricular organs, takes place. These will, via ascending projections, ultimately activate corticotropin-releasing hormone and arginine vasopressin releasing neurons in the paraventricular nucleus of the hypothalamus that control the release of adrenocorticotropin hormone in the anterior pituitary and, in turn, the release of corticosteroids in the adrenal cortex. If the stressor is a psychosocial stimulus, activation may occur in the amygdala, hippocampus and/or frontal cortex, among other limbic brain structures that modulate the activity of distinct nuclei in the bed nucleus of the stria terminalis or of the nucleus of the solitary tract, dorsomedial hypothalamic nucleus, arcuate nucleus or peri-paraventricular nucleus zone and, subsequently, the activity of the paraventricular nucleus (Sousa, 2016). These facts provide better insight into DU role as a neuroendocrine disruptor acting at all levels of integrative regulatory pathways in the body.

We have already mentioned a reciprocal relationship between low-dose rates of ionizing radiation and their harmful effects on the cell. Biological effects are extensive and unpredictable, because of their complex regulation and the fact that they are dependent on exposition, as well as on genetic predisposition, health, age and other individual factors (Zunic and Rakic, 2015).

Literature data showed that the effects of insoluble plutonium dioxide ($^{239}PuO_2$) aerosols (α-particles emitter), which were inhaled in experimental conditions, resulted in primary no carcinogenic changes (lymphopenia, atrophy and fibrosis of the thoracic lymph nodes, radiation pneumonitis and pulmonary fibrosis), which preceded delayed cancer lesions (Muggenburg et al., 2008). This study highlighted the importance of distribution of inhaled α-emitting radionuclides. A more uniform distribution of α-particle dose within the lung poses a greater risk of neoplasia, than less uniform distributions of α-particle dose.

As cancer was detected earlier, patient's survival would be longer (Midthun and Jett, 2009). The authors speculated that for aggressive cancer starting from a single cell, death may be achieved in 5 years, and symptoms may be detected very late, in year 4. Screening prolongs survival. The chest radiograph is a routine procedure, easy to perform, and well accepted by patients, but it is not sensitive enough as computerized tomography (CT) scan (Midthun and Jett, 2009). This study summarized a prevalence of nodules and cancers detected in a prospective, single-arm observational low-dose spiral CT screening of ten studies. Out of 53,399 subjects included, 628 cases of lung cancers were detected, out of which 510.25 in surgical stage I A/B.

In consonance with the slogan - every life counts, the use of CT scan as a screening tool is highly rational! But, a study by Mathews and coworkers (2013) explained that the overall cancer incidence was 24% greater for exposed than for unexposed people. The mean duration of follow-up was 9.5 years after exposure.

One more reason which contributes to the misunderstanding of DU radio-genotoxic and overall biohazard is the high incidence of lung cancer in cigarette smokers. There have been 20 million tobacco-related deaths since 1964 (CDC, 2008). After embedding into the tissue, DU micro or nanoparticles exert the heavy metal toxic effects, as well as the radiogenotoxic effects of mixed, alpha, beta and gamma radiation. Nanoparticles can produce irreversible damage to cells by oxidative stress or/and organelle injury, preceding tissue inflammation and altered cell death mechanisms. The carcinogenic effect of tobacco smoke may act synergically with tobacco-containing radionuclides, which are mostly alpha-emitters which induce cumulative doses at bifurcations (Martel, 1983). No matter whether lung cancer developed as a result of smoking habit, or due to internal contamination with radionuclides from the environment or from tobacco, the disease may occur many years after the intake, because of the lag period between the damage of sensitive cells and the appearance of recognizable tumors (Priest, 2001)…

Our results (Zunic et al., 2007) are in relation with a publication authored by Smith et al., (2015), who proposed that lung cancer screening procedure should be performed with heavy smokers and ex-smokers (within 15 years after smoking cessation). In our pilot study, all patients with lung cancer were heavy smokers with smoking history >30 pack-years. This is the reason why we believe that all smokers who undergo bronchology unit should be included in the fast screening procedure, which we presented in this study as an initial step in clinical examination. Simplicity and low cost are advantages of this method (Zunic, 2015).

A study authored by Mihailovic and coworkers (2013), showed the increasing trends in both overall cancer incidence and mortality rates which were identified for Serbia (Figure 19).

According to published data (Mihailovic et al., 2013), in men, lung cancer showed the highest incidence, followed by colorectal, prostate and bladder cancer. Breast cancer was the most common form of cancer in women, followed by cervical, colorectal and lung cancer. Prostate and colorectal cancer incidence has been significantly increasing over the last years in men, while this was also observed for breast cancer incidence and lung cancer mortality in women.

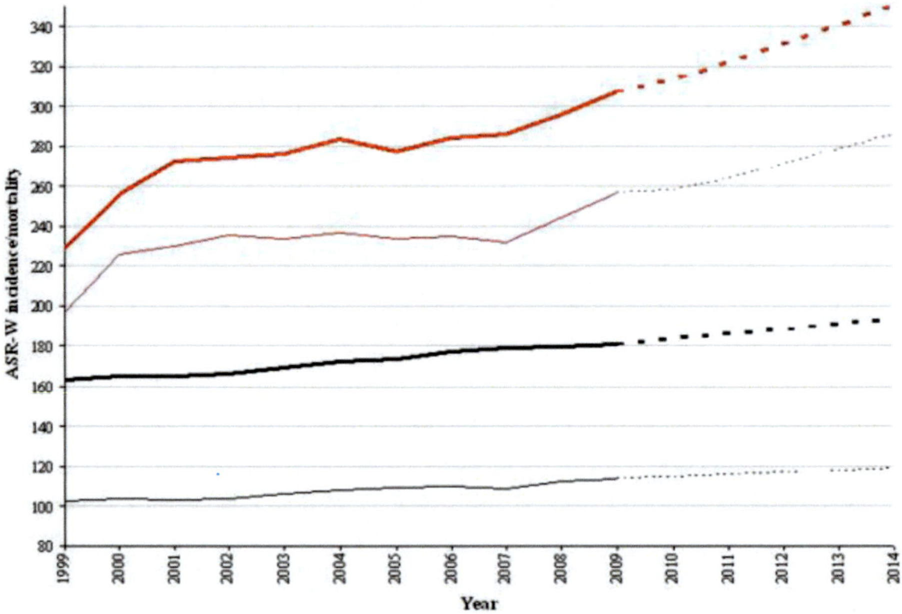

Figure 19. The overall cancer incidence and mortality in Serbia - trends (1999–2009) and predictions (2010–2014).
Legend: The red bold line – incidence, men. The red regular line – incidence, women. The black bold line – mortality, men. The black regular line - mortality, women. All dotted lines continuous to these are depicting predictions. From: Mihailovic, J; Pechlivanoglou, P; Miladinov-Mikov, M; Živković, S; Postma, MJ. Cancer incidence and mortality in Serbia 1999–2009. *BMC Cancer*, 2013;13:18. DOI: 10.1186/1471-2407-13-18. This article is published under the license of BioMed Central Ltd. This is an Open Access article distributed under the terms of the Creative Commons Attribution License (http://creativecommons.org/licenses/by/2.0), which permits unrestricted use, distribution, and reproduction in any medium, provided that the original work is properly cited.

Cancer incidence and mortality in Serbia has been generally increasing over the 10-year period since the bombing of Serbia and Montenegro. These results, reported by Mihailovic et al., (2013), speak in favor of our understanding of the early and delayed health effects of DU. Apparently, the increasing trend of the overall cancer incidence in Serbia started immediately after the bombing of Serbia and nearby territory with DU projectiles. This might be a consequence of prolonged (since 1990) exposure to long-lived isotopes, which were released into the air in the Persian Gulf and Bosnia and Herzegovina (1994/5). Other possibility is that stress in the population, which was exhausted by the EU sanctions, bombing, social conflicts, supported the expression of some kind of maladaptation (according to Sousa, 2016), which contributed to alarmingly higher cancer incidence and mortality in Serbia than in the majority of European regions (Mihailovic et al., 2013).

The possibility to "measure radiation hormesis" should be one of the key strategies in lung cancer screening. The method was based on the hormesis-threshold model of tissue response to low-dose radiation, especially from alpha-emitting radionuclides, which is a step forward in achieving individualized screening and lung cancer risk assessment (Zunic, 2015; Zunic and Rakic, 2015). The proposed model does not require an exact measuring of tissue doses in conditions of exposure to low doses of radiation, especially alpha emitting radioisotopes. This method enables a quick orientation concerning the extent of damage of complex tissue regulatory mechanisms in situ, and indirectly, it may indicate the existence of adaptive, premalignant or malignant lesions (Zunic, 2015).

An additional benefit may be achieved from the analysis of metabolome outputs (Zunic, 1993; Zunic et al., 2007). Carracedo, Cantley and Pandolfi, (2013) discussed the malfunction of mitochondria as a leading contributing factor which triggers the alterations in metabolism. We have already discussed in this study the changes of metabolome under low-dose radiation exposure. There was detected a causal relationship between cancer genes and metabolic alterations, and their potential to be targeted for cancer treatment (Carracedo, Cantley and Pandolfi, 2013).

We propose the estimation of hormetic tissue response to carcinogenic stimuli to be the first step in lung cancer screening procedure. The advantage of this approach is a delay, or avoiding of potentially unnecessary imaging procedures, which are based on radiation impact. The next one may include conventional cytogenetic testing methods, as well as comprehensive gene expression analysis, with the aim to elaborate the earliest changes induced by the noxious agent, intoxicants, and/or irradiation.

Sousa (2016) has reported that there has been an increasing interest in epigenetic mechanisms in the past two decades. Early-life stress events trigger a developmental deregulation of epigenetic pathways that result in discrete or genome-wide changes in gene expression in various tissues, including the brain. These subtle changes may be a subject of research in the field of advanced biotechnology. A remarkable variability in the individual response and predisposition to the effects of stress has a multifactorial origin, but genetic and epigenetic mechanisms are implicated in it. Genetically transmitted patterns of reaction to stressors are highly preserved within species because they are critical for survival and evolution (Sousa, 2016).

Having in mind all these facts, we conclude that radiation has, almost since its discovery, about 120 years ago, been used not only to provide energy, or for medical purposes. Nowadays, radiation is the most powerful weapon, the ideal, invisible killer, which, in case of the military use of depleted uranium, as we discussed in this paper, has already irreversibly changed all natural resources, contributed to mass migration of population, destruction of social relations, and *in vivo* experimentation with the health of human population and the overall living world.

REFERENCES

Abbasi, R; Abu-Zayyad, T; Belov, K; Belz, J; Cao, Z; Dalton, M; Fedorova, Y; Huentemeyer, P; Jones, BF; Jui, CCH; Loh, EC; Manago, N; Martens, K; Matthews, JN; Maestas, M; Smith, J; Sokolsky, P; Springer, RW; Thomas, J; Thomas, S; Chen, P; Field, C; Hast, C; Iverson, R; Ng, JST; Odian, A; Reil, K; Walz, D; Bergman, DR; Thomson, G; Zech, A; Chang, FY; Chen, CC; Chen, CW; Huang, MA; Hwang, WYP; Lin, GL. Air fluorescence measurements in the spectral range 300-420 nm using a 28.5 GeV electron beam. *Astropart Phys,* 2008; 29:77–86.

Acheva, A; Aerts, A; Rombouts, C; Baatout, S; Salomaa, S; Manda, K; Kamarainen, M. Human 3D tissue models in radiation biology: current status and future perspectives. *Int J Radiat Res*, 2014; 12: 81-98.

AGNIR, 2004. Advisory Group on Non-ionising Radiation (AGNIR). [2013 May 24]. Available at: http://webarchive.nationalarchives.gov.uk/20140629102627/http:/www.hpa.org.uk/Topics/Radiation/RadiationAdvisoryGroups/AdvisoryGroupOnNonIonisingRadiation/.

Al-Muqdadi, K; Al-Ansari, N. [Depleted Uranium: Its nature, characteristics and risks of the Military uses on Humans and the Environment]. [2009 February 12]. Available at: http://www.ltu.se/cms_fs/1.85152!/ file/4.3%20Almuqdadi%20and%20alansari%202.pdf.

Allison, I; Bindoff, N; Bindschadler, R; Cox, P; de Noblet-Ducoudre, N; England, M; Francis, J; Gruber, N; Haywood, A; Karoly, D; Kaser, G; Le Quéré, C; Lenton, T; Mann, M; McNeil, B; Pitman, A; Rahmstorf, S; Rignot, E; Schellnhuber, HJ; Schneider, S; Sherwood, S; Somerville, R; Steffen, K; Steig, E; Visbeck, M; Weaver, A. [The Copenhagen Diagnosis, 2009: Updating the World on the Latest Climate Science]. [2009 November], by UNSW Climate Change Research Centre, Sydney.

Available at: http://www.eea.europa.eu/data-and-maps/figures/number-of-reported-climate-related.

Amundson, SA; Lee, RA; Koch-Paiz, CA; Bittner, ML; Meltzer, P; Trent, JM; Fornace, AJ. Differential responses of stress genes to low dose-rate gamma irradiation. *Mol Cancer Res*, 2003; 1(6):445-52.

Antanaitis, BC; Aisen, P. Uteroferrin and the purple acid phosphatases. *Adv Inorg Biochem*, 1983; 5:111-36.

Aoshiba, K; Tamaoki, J; Nagai, A. Acute cigarette smoke exposure induces apoptosis of alveolar macrophages. *Am J Physiol Lung Cell Mol Physiol,* 2001; 281:L1392–L1401.

ASRC. [Radiation compromises DNA repair depending on its LET]. In: *DNA Damage and Radioactive Materials.* Japan Atomic Energy Agency. http://asrc.jaea.go.jp/soshiki/gr/eng/mysite6/sub3.html. (2016 April 29]. Available at: http://asrc.jaea.go.jp/soshiki/gr/eng/mysite6/sub3.html

Averill, BA; Davis, JC; Burman, S; Zirino, T; Sanders-Loehr, J; Loehr, TM; Sage, JT; Debrunner, PG. Spectroscopic and magnetic studies of the purple acid phosphatase from bovine spleen. *J Am Chem Soc*, 1987; 109:3760–7.

Azimian, H; Bahreyni-Toossi, MT; Rezaei, AR; Rafatpanah, H; Hamzehloei, T; Fardid, R. Up-regulation of Bcl-2 expression in cultured human lymphocytes after exposure to low doses of gamma radiation. *Journal of Medical Physics/Association of Medical Physicists of India*, 2015; 40(1):38-44. doi:10.4103/0971-6203.152249.

Bachelor, PP; Jordan, DV; Harper, WW; Cannon, BD, Finn, EC. Self-absorption effects on alpha-induced atmospheric nitrogen fluorescence yield. *Journal of radioanalytical and nuclear chemistry,* 2009; 282:873-6. DOI 10.1007/s10967-009-0331-0.

Barnes, EW. Basic physics of nuclear medicine. In: *Nuclear medicine.* Henkin, RE; Boles, MA; Dillehay, GL; Halama, JR; Karesh, SM; Wagner, RH; Zimmer, MA. (ed.). Mosby-Year Book St. Louis Baltimore Boston Carlsabad Chicago Naples New York Philadelphia Portland London Madrid Mexico City Singapore Sydney Tokyo Toronto Wiesbaden. 1996; pp. 43-63.

Baverstock, K. Mutation Research/Fundamental and Molecular Mechanisms of Mutagenesis. *Mutat Res*, 2000; 454(1-2):89-109. doi:10.1016/S0027-5107(00)00100-7.

Baverstock, FK. [The toxicity of Depleted uranium]. Presentation to the Defence Committee of the Belgian House of Representatives: [2006 November 20]. Available at: http://www.kbaverstock.org/page6.html.

Beck, JL; Keough, DT; De Jersey, J; Zerner, B. Enzymatically active zinc, copper and mercury derivatives of the one-iron form of pig allantoic fluid acid phosphatase. *Biochim Biophys Acta*, 1984; 791(3):357-63.

Behari, J. Biological Effects and Health Implication of Radiofrequency and Microwave. International Conference on Electromagnetic Interference and Compatibility. *Conference Procedings*. IEEE: New Delhi, India, 1999; 449-52.

Belyakov, OV; Folkard, M; Mothersill, C; Prise, KM; Michael, BD. Bystander-induced apoptosis and premature differentiation in primary urothelial explants after charged particle microbeam irradiation. *Radiat Prot Dosimetry*, 2002; 99:249-251.

Bertell, R. Depleted uranium: all the questions about DU and Gulf War syndrome are not yet answered. *Int J Health Serv,* 2006; 36:503–20.

Benton, C. [Alpha particle]. [2015 June 15, 19:09]. Available at: https://bentonaerospace.wordpress.com/2012/01/09/alpha-particle/.

Betti, M. Civil use of depleted uranium. *J Environ Radioact*, 2003; 64:113–9.

Bevilacqua, MA; Lord, DK; Cross, NC; Whitaker, KB; Moss, DW; Cox, TM. Regulation and expression of type V (tartarate-resistant) acid phosphatase in human mononuclear phagocytes. *Mol Biol Med*, 1991; 8(1):135-40.

Blanchard, JP; Blackman, CF. Clarification and amplification of an ion parametric resonance model for magnetic field interactions with biological systems. *Bioelectromagnetics*, 1994; 15:217-238.

Blank, M; Goodman, R. Electromagnetic fields stress living cells. *Pathophysiology,* 2009; 16(2-3):71-8. doi: 10.1016/j.pathophys. 2009. 01.006.

Bleise, A; Danesi, PR; Burkart, W. Properties, use and health effects of depleted uranium (DU): a general overview. *J Environ Radioact,* 2003; 64:93–112.

Bohr, VA; Phillips, DH; Hanawalt, PC. Heterogeneous DNA damage and repair in the mammalian genome. *Cancer Res*, 1987; 47:6426-36.

Bonner, WM. Low-dose radiation: Thresholds, bystander effects, and adaptive responses. *PNAS,* 2003; 100(9):4973–5. Available at: www.pnas.org _cgi_doi_10.1073_pnas.1031538100.

Borthacur, GG. [Is there a safe level of radiation exposure? –The Petkau effect]. Available at: http://physics.unipune.ernet.in/~phyed/27.4/1384pdf. pdf. (2016 April 24).

Bowen, R. [Mechanism of Action: Hormones with Intracellular Receptors]. [1998 May 27]. Available at: http://arbl.cvmbs.colostate.edu/hbooks/pathphys/endocrine/moaction/intracell.html.

Brack, HG. Fukushima Daiichi Nuclear Information Handbook. In: *A Guide to Accident Terminology and Information Sources*. Pennywheel Press, Hulls Cove. 2011. Available at: http://www. davistownmuseum. org/PDFs /Nuclear%20Disaster%20in%20Japan_Draft.pdf.

Bradford, PT; Freedman, MD; Goldstein, AM; Tucker, MA. Increased Risk of Second Primary Cancers After a Diagnosis of Melanoma. *Archives of Dermatology*, 2010; 146(3):265-272. doi:10.1001/archdermatol.2010.2.

Branchfield, K; Nantie, L; Verheyden, JM; Sui, P; Wienhold, MD; Sun, X. Pulmonary neuroendocrine cells function as airway sensors to control lung immune response. *Science*, 2016; DOI: 10.1126/science.aad7969.

Burger, M. The risks of depleted uranium contamination in post conflict countries: Findings and lessons learned from UNEP field assessments. In: *Assessing and Restoring Natural Resources in Post Conflict Peace-building*. Jensen, D; Lonergan, S. (ed). Earthscan, London, 2012. Available at: http://environmentalpeacebuilding.org/sites/default/ files /vol.5_ch.8_burger_colorcoverpage.pdf.

Burlakova, EB; Goloshchapov, AN; Zhizhina, GP; Konradov, AA. New aspects of the effects of low intensity radiation. *Radiats Biol Radioecol*, 1999; 39(1):26-34.

Burstone, MS. Histochemical demonstration of acid phosphatase activity in osteoclasts. *J Histochem Cytochem*, 1959; 7:39-41.

Busby, C. Aspects of DNA Damage from Internal Radionuclides. Chapter 22. In: *New Research Directions in DNA Repair*. Chen, C. (ed.). InTech, 2013; 597-637. Available at: http://dx.doi.org/10.5772/53942.

Busby, C; Morgan, S. [Did the Use of Uranium Weapons in Gulf War 2 Result in Contamination of Europe? Evidence from the Measurements of the Atomic Weapons Establishment] [2006] Aldermaston, Aberystwyth, Green Audit. Available at: http://www.llrc.org/aldermastrept.pdf.

Busby, C; Hamdan, M; Ariabi, E. Cancer, infant mortality and birth sex-ratio in Fallujah, Iraq 2005–2009. *International journal of environmental research and public health*, 2010; 7(7):2828-37.

Buzea, C; Blandino, IIP; Robbie, K. Nanomaterials and nanoparticles: sources and toxicity. *Biointerphases*, 2007; 2(4):MR17-MR71.

Cacciapaglia, F; Spadaccio, C; Chello, M; Gigante, A; Coccia, R; Afeltra, A; Amoroso, A. Apoptotic molecular mechanisms implicated in autoimmune diseases. *Eur Rev Med Pharmacol Sci*, 2009; 13(1):23-40.

Carracedo, A; Cantley, LC; Pandolfi, PP. Cancer metabolism: fatty acid oxidation in the limelight. *Nature Reviews Cancer, 2013;* 13:227-232. doi:10.1038/nrc3483.

CDC, 2008. Centers for Disease Control and Prevention. Smoking-attributable mortality, years of potential life lost, and productivity losses - United States, 2000-2004. *Morb Mortal Wkly Rep*, 2008; 57:1226-8. Available at: http://www.cdc.gov/mmwr/preview/mmwrhtml/mm5745a3.htm.

Chinnery, PF; Turnbull, DM. Mitochondrial DNA and disease. *The Lancet,* 1999; 354(9176): si17–si21. doi:10.1016/S0140-6736(99)90244-1.

Chmyrev, V; Isaev, N; Bilichenko, S; Stanev, G. Observation by space-borne detectors of electric fields and hydromagnetic waves in the ionosphere over an earthquake centre. *Phys Earth Planet Inter*, 1989; 57(1):110–4.

Clark, De-L. [U235: A Gamma Ray Analysis Code for Uranium Isotopic Determination]. [1996 December] Work performed under the auspices of the US. Department of Energy by the Lawrence Livermore National Laboratory under Contract W-7405-ENG-48.1996. DOI: 10.2172/310887. Available at: http://www.osti.gov/scitech/servlets/purl/310887/. (2016 April 26).

Coates, PJ; Lorimore; SA; Wright, EG. Damaging and protective cell signalling in the untargeted effects of ionizing radiation. *Mutat Res*, 2004; 568(1-2):5-20. doi:10.1016/j.mrfmmm.2004.06.042.

Cohen, PL; Caricchio, R; Abraham, V; Camenisch, TD; Jennette, JC; Roubey, RAS; Earp, HS; Matsushima, G; Reap, EA. Delayed Apoptotic Cell Clearance and Lupus-like Autoimmunity in Mice Lacking the c-mer Membrane Tyrosine Kinase. *J Exp Med*, 2002; 196(1):135–140.

Cohen, S; Houben, A; Segal, D. Extrachromosomal circular DNA derived from tandemly repeated genomic sequences in plants. *The Plant Journal,* 2008; 53(6):1027–34. doi:10.1111/j.1365-313X.2007.03394.x.

Cohen, S; Agmon, N; Sobol, O; Segal, D. Extrachromosomal circles of satellite repeats and 5S ribosomal DNA in human cells. *Mobile DNA,* 2010; 1(1):1–11. doi:10.1186/1759-8753-1-11.

Coleman, MA; Yin, E; Peterson, LE; Nelson, D; Sorensen, K; Tucker, JD; Wyrobek, AJ. Low-dose irradiation alters the transcript profiles of human lymphoblastoid cells including genes associated with cytogenetic radioadaptive response. *Radiat Res,* 2005; 164:369-82.

Cvetkovic, P. Values of cytochemical examination of normal and pathologic T, B and "O" lymphocytes. *Doctoral thesis*. Belgrade University, Medical Faculty, 1981.

Devi, JJ; Gupta, T; Rajmal, J; Tripathi, SN. Measurement of personal and integrated exposure to particulate matter and co-pollutant gases. *Environmental Science and Pollution Research*, 2012; 20:1632-48. doi:10. 1007/s11356-012-1179-3.

Djordjevic-Denic, G; Radak, Dj; Cvetkovic, P; Perovic, M. Cytochemical examination of the amount of lipids in alveolar macrophages during experimental arteriosclerosis. *Iugoslav Physiol Pharmacol Acta*, 1987; 23(1):105-6.

Djordjevic, D; Buha, J; Stortini, AM; Mihajlidi-Zelic, A; Relic, D; Barbante, C; Gambaro, A. Mass distributions and morphological and chemical characterization of urban aerosols in the continental Balkan area (Belgrade). *Environmental Science and Pollution Research,* 2016; 23(1):851-9. DOI 10.1007/s11356-015-5271-3

Djurovic, B; Selakovic, V; Spasic-Jokic, V. Does occupational exposure to low-dose ionizing radiation induce cell membrane damage? *Arch Oncol*, 2004; 12(4):197-9.

Djurovic, B; Spasic-Jokic, V; Djurovic, B. Influence of occupational exposure to low-dose ionizing radiation on the plasma activity of superoxide dismutase and glutathione level. *Vojnosanit Pregl*, 2008; 65(8):613-8.

DOE-HDBK (US Department of Energy Handbook), [Primer on Spontaneous Heating and Pyrophoricity, Chapter: Uranium] [1994 December]. Available at: http://energy.gov/sites/prod/files/2013/06/f2/hdbk1081.pdf. (2016 April 26).

Doi, K; Antanaitis, BC; Aisen, P. The binuclear iron centers of uteroferrin and the purple acid phosphatases. *Struct Bond*, 1988; 70:1-26.

Durakovic, A. Medical Effects of Internal Contamination with Uranium. *CMJ,* 1999; 40(1):1-18. Available at: http://www.mindfully.org/Nucs/DU-Medical-Effects-Mar99.htm/. (2016 April 26).

EEA, 2012. EEA Report No 12/2012. [Climate change, impacts and vulnerability in Europe 2012. An indicator-based report. Number of reported extreme weather events and wildfire]. European Environment Agency [Last modified: 2012 Nov 30, 10:50 AM], http://www.eea.europa.eu/publications/climate-impacts-and-vulnerability-2012. (2016 April 26].

Elgazzar, AH; Kazem, N. Biological Effects of Ionizing Radiation. Chapter 23. In: *The pathophysiologic basis of nuclear medicine*. Elgazzar, AH. (ed.). Springer Science & Business Media, 2006.

Ellsworth, WL. Injection-Induced Earthquakes. *Science*, 2013; 341(6142): 1225942. doi: 10.1126/science.1225942.

EM-DAT, 2012. [Number of reported extreme weather events and wildfire]. European Environment Agency. Emergency Events Database. [2012 Nov 30]. Available at: http://www.eea.europa.eu/data-and-maps/data/external/emergency-events-database-em-dat.

Fachin, AL; Mello, SS; Sandrin-Garcia, P; Junta, CM; Ghilardi-Netto, T; Donadi, EA; Passos, GA; Sakamoto-Hojo, ET. Gene expression profiles in radiation workers occupationally exposed to ionizing radiation. *J Radiat Res*, 2009; 50(1):61-71.

Fenech, M; Kirsch-Volders, M; Natarajan, AT; Surralles, J; Crott, JW; Parry, J; Norppa, H; Eastmond, DA; Tucker, JD; Thomas, P. Molecular mechanisms of micronucleus, nucleoplasmic bridge and nuclear bud formation in mammalian and human cells. *Mutagenesis*, 2011; 26(1):125-32.

Fireman, P. *Atlas of allergies: Immunology of Allergic Disorders*. Fireman, P; Slavin, RG. (ed.). Mosby-Wolfe, an imprint of Times Mirror International Publishers Limited,London, Baltimore, Barcelona, Bogota Boston, Buenos Aires, Caracas, Carlsbad, CA, Chicago, Madrid, Mexico City, Milan, Naples, FL, New York, Philadelphia, St. Louis, Seoul, Singapoore, Sydney, Taipei, Tokyo, Toronto, Wiesbaden, 1996; 1-26.

Fleckenstein, E; Dirks, W; Dehmel, U; Drexler, HG. Cloning and characterization of the human tartrate-resistant acid phosphatase (TRAP) gene. *Leukemia*, 1996; 10(4):637-43.

Florence, TJ; Reiser, MB. Hot on the trail of temperature processing. *Nature,* 2015; 519:296-7. DOI: http://dx.doi.org/10.1038/nature14209.

Formica, D; Silvestri, S. Biological effects of exposure to magnetic resonance imaging: an overview. *BioMedical Engineering OnLine*, 2004; 3:11. doi:10.1186/1475-925X-3-11.

Fox, KC. [NASA's Goddard Space Flight Center, Greenbelt, Md. Lightning-made Waves in Earth's Atmosphere Leak Into Space]. [2011 November 28]. Available at: http://www.nasa.gov/mission_pages/sunearth/ news /lightning-waves.html. (2016 April 24)

Fraga, M M; Onofre, A; Pereira, L; Castro, N; Veloso, F; Fraga, F; Ferreira Marques, R; Pimenta, M; Policarpo, A. Temperature-dependent quenching of UV fluorescence of N_2. *Nucl Instrum Methods Phys Res*, 2008; 597:75-82. DOI: 10.1016/j.nima.2008.08.046.

Furlong, H; Mothersill, C; Lyng, FM; Howe, O. Apoptosis is signalled early by low doses of ionizing radiation in a radiation-induced bystander effect. *Mutat Res*, 2013; 741:35-43. [PubMed: 23454491].

Gatti, M; Montanari, S. [Impact on health by nanoparticles created by high temperature explosions]. In: 8th ETH Conference on Combustion Generated Nanoparticles. [2004 August]. Available at: http://citeseerx. ist.psu.edu/viewdoc/download?doi=10.1.1.297.1017&rep=rep1&type=pdf .

Gehr, P; Heyder, J. eds. *Particle-lung interactions*. CRC Press, 2000.

Griffin, DW. Atmospheric Movement of Microorganisms in Clouds of Desert Dust and Implications for Human Health. *Clin Microbiol Rev*, 2007; 20(3):459-77. doi:10.1128/CMR.00039-06.

Große, L; Wurm, CA; Brüser, C; Neumann, D; Jans, DC; Jakobs, S. Bax assembles into large ring-like structures remodeling the mitochondrial outer membrane in apoptosis. *The EMBO Journal,* 2016; 35:402–13. DOI 10.15252/embj.201592789.

Gulhane, S; Gangane, N. Detection of lupus erythematosus cells in pleural effusion: An unusual presentation of systemic lupus erythematosus. *J Cytol*, 2012; 29(1):77-9. doi: 10.4103/0970-9371.93232.

Gupta, TK. Nuclear radiation, ionization and reactivity. In: *Radiation, Ionization, and Detection in Nuclear Medicine*. (ed.). Springer Heidelberg New York Dordrecht London. 2013; pp. 1-58.

Haaf, T; Raderschall, E; Reddy, G; Ward, DC; Radding, CM; Golub, EI. Sequestration of mammalian Rad51-recombination protein into micronuclei. *J Cell Biol*, 1999; 144:11–20.

Hainsworth, LB. The Effect of Geophysical Phenomena on Human Health. *Speculations in Science and Technology*, 1983; 6(5):439-44.

Hall, EJ; Hei, TK. Genomic instability and bystander effects induced by high-LET radiation. *Oncogene*, 2003; 22(45):7034-42.

Halleen, JM; Jaija, H; Stepan, JJ; Vihko, P; Väänänen, HK. Studies on the tyrosine phosphatase activity of tartarate-resistant acid phosphatase. *Arch Biochem Biophys*, 1998; 352:97-102.

Halleen, J; Saisanen, S; Salo, J; Reddy, SV; Roodman, GD; Hentunen, TA; Lehenkari, PP; Kaija, H; Vihko, P; Väänänen, HK. Intracellular fragmentation of bone resorption products by reactive oxygen species generated by osteoclastic tartarate-resistanmt acid phosphatase. *J Biol Chem*, 1999; 274: 22907-10.

Halleen, J. Tartrate-resistant Acid Phosphatase 5b (TRACP 5b) as a Marker of Bone Resorption. *Rev Ser*, 2006; 3:1-10.

Hand, L. [CT Scans May Increase Cancer Risk in Children, Adolescents]. [2013 May 23]. *Medscape*. Available at: http://www.medscape.com/viewarticle/804715.

Harmon, BV; Winterford, CM; O'Brien, BA; Allan DJ. Morphological criteria for identifying apoptosis. In: *Cell Biology*. A Laboratory Handbook, Second Edition 1. Eds Academic Press: New York, 1988; 327-40.

Haslam, P; Watling, A; Evans, P; et al. Training Course in Bronchoalveolar Lavage Handbook of Techniques and Protocols. 1984.

Haugen, A; Ryberg, D; Mollerup, S; Zienolddiny, S; Skaug, V; Svendsrud, DH. Gene environment interactions in human lung cancer. *Toxicol Lett*, 2000; 112-113:233-7.

Hayhoe, F; Quaglino, D. Hematological cytochemistry. Churchill Livingstone, London, 1980.

Hayman, AR; Cox, TM. Purple acid phosphatase of the human macrophage and osteoclast. *J Biol Chem*, 1994; 269:1294-300.

Heerdt, BG; Houston, MA; Augenlicht, LH. Potentiation by specific short-chain fatty acids of differentiation and apoptosis in human colonic carcinoma cell lines. *Cancer Res*, 1994; 54:3288-94.

Hepburn, AL. Heberden Historical Series. The LE cell. *Rheumatology,* 2001; 40(7):826-827. doi:10.1093/rheumatology/40.7.826.

Hittelman, VN. Genetic Instability in Epithelial Tissues at Risk for Cancer *Ann N Y Acad Sci,* 2001; 952:1-12.

Hoff, O. Determination of Alpha Radiation-Induced Fluorescence Efficiency. *Honors Theses. Paper 38*, 2012. Available at: http://aquila.usm.edu/cgi/viewcontent.cgi?article=1037&context=honors_theses.

Hohenemser, C; Deicher, M; Hofsass, H; Lindner, G; Recknagel, E; ; Budnick, J. Agricultural impact of Chernobyl: a warning. *Nature,* 1986; 321:817. doi:10.1038/321817a0.

Hughes, D; Haslam, LP; Townsend, D; Turner-Warwick, M. Numerical and functional alterations in circulatory lymphocytes in cigarette smokers. *Clin Exp Immunol*, 1985; 61:459-66.

Hughes, DA; Haslam, PL. Effect of smoking on the lipid composition of lung lining fluid and relationship between immunostimulatory lipids, inflammatory cells and foamy macrophages in extrinsic allergic alveolitis. *Eur Respir J*, 1990; 10:1128-39.

IAEA, 1997. IAEA-TECDOC-934. Effects of ionizing radiation on blood and blood components: A Survey IAEA, Vienna, 1997.

IARC, 2012. Radiation. IARC Monographs on The Evaluation of Carcinogenic Risks to Humans. World Health Organization, International Agency for Research on Cancer, 2012; 100(D):7–303.

IGMASS, 2012: [Towards International Collaboration in the Defense of Mankind] Conference Report by Benjamin Deniston, Pavel Penev, and Jason Ross, Space and Global Security of the Humanity IV International Specialized Symposium, Yevpatoria, Ukraine [2012 September 3-12]. Available at: http://larouchepac.com/node/23994.

Il'yasova, D; Kinev, A; Melton, CD; Davis, FG. Donor-specific cell-based assays in studying sensitivity to low-dose radiation: a population-based perspective. *Front Public Health,* 2014; 2:244. doi: 10.3389/fpubh.2014.00244. eCollection 2014.

Inan, OT. [Interactions of Electromagnetic Waves with Biological Tissue]. *BIOE 200C.* Lecture. [2005]. Avalable at: http://www.emraware.com/Documents/class_12.pdf.

Inoue, K; Takano, H. Aggravating impact of nanoparticles on immune-mediated pulmonary inflammation. *Scientific World Journal,* 2011; 11:382-90. doi: 10.1100/tsw.2011.44.

IPPNW, 2014. IPPNW Report. The health effects of uranium weapons. The social debate on the use of a controversial weapon. 1st edition. August 2014.

IPPNW, 2015. International Physicians for the Prevention of Nuclear War, European Conference, Belgrade, 11-13 September 2015.

Israeli, E. Gulf War syndrome as a part of the autoimmune (autoinflammatory) syndrome induced by adjuvant (ASIA). *Lupus*, 2012; 21(2):190-4.

Iyer, R; Lehnert, BE. Alpha-particle-induced increases in the radioresistance of normal human bystander cells. *Radiation Res*, 2002; 157(1):3-7.

Jäättelä, M; Tschopp, J. Caspase-independent cell death in T lymphocytes. *Nat Immunol*, 2003; 4(5):416-23.

Jaubert, F; Monnet, JP; Danel, C; Chretien, J; Nezelof, C. The location of non-specific esterase in human lung macrophages. An ultrastructural study. *Histochemistry*, 1978; 59(2):141-7.

Jevremovic, T. Beta decay. Mechanism of beta decay. In: *Nuclear Principles in Engineering.* Springer Science & Business Media. (ed.). New York 2009; pp. 193-7. ISBN 978-0-387-85608-7.

Jin, YW; Na, YJ; Lee, YJ; An, S; Lee, JE; Jung, M; Kim, H; Nam, SY; Kim, CS; Yang, KH; Kim, SU; Kim, WK; Park, WY; Yoo, KY; Kim, CS; Kim, JH. Comprehensive analysis of time- and dose-dependent patterns of gene expression in a human mesenchymal stem cell line exposed to low-dose ionizing radiation. *Oncol Rep*, 2008; 9(1):135-44.

Jirtle, RL; Skinner, MK. Environmental epigenomics and disease susceptibility. *Nat Rev Genet,* 2007; 8:253-62.

Johnson, N; Haslam, LP; Dewar, A; Newman-Taylor, A; Turner-Warwick, M. Identification of inorganic dust particals in bronchoalveolar lavage macrophages by energy dispersive x-ray microanalysis. *Arch Environ Health*, 1986; 41(13):133-144.

Joshi, T; Graham, B; Spiccia, L. Macrocyclic Metal Complexes for Metalloenzyme Mimicry and Sensor Development. *Acc Chem Res*, 2015; 48(8):2366-79. DOI: 10.1021/acs.accounts.5b00142.

Jovanovic, V; Oldja, M. [Video control of wildfires in Serbia. Ministry of Agriculture, Forestry and Water Management]. Directorate of Forests, Belgrade Serbia and Public Enterprise “Vojvodinasume. [2007]. Available at: http://www.fire.uni-freiburg.de/sevilla-2007/contributions/doc/SESIONES_TEMATICAS/ST7/Jovanovic-Oldja_SERBIA.pdf.

Kaatsch, P; Spix, C; Schulze-Rath, R; Schmiedel, S; Blettner, M. Leukaemia in young children living in the vicinity of German nuclear power plants. *Int J Cancer*, 2008; 1220:721-6.

Keller, M; Anet, B, Burger M, Schmid E, Wicki A, Wirz C. [Depleted uranium: Environmental and health effects in the Gulf War, Bosnia and Kosovo]. European Parliament, Luxembourg. Directorate-General for Research. Division for Industry, Research, Energy, Environment and STOA. [2001 May]. Available at: http://www.uni-mannheim.de/edz/pdf/dg4/STOA100_EN.pdf.

Kim, YH; Park, YJW; Lee, JY; Kwon, TK. Sodium butyrate sensitizes TRAIL-mediated apoptosis by induction of transcription from the DR5 gene promoter through Sp1 sites in colon cancer cells. *Carcinogenesis,* 2004; 25(10):1813-20. doi: 10.1093/carcin/bgh188.

Kim, C; Lee, O; Kim, JY; Sung, W; Lee, NK. Dynamic Release of Bending Stress in Short dsDNA by Formation of a Kink and Forks. *Angew Chem Int*, 2015; 54:8943-7. doi:10.1002/anie.201502055.

Koban, A; Geipel, G; Roßberg, A; Bernhard G. Uranium (VI) complexes with sugar phosphates in aqueous solution. *Radiochimica Acta/International journal for chemical aspects of nuclear science and technology*, 2004; 92(12/2004):903-8.

Koch, J; Vogt, G; Kissel, W. Cytoplasmic DNA is structurally different from nuclear DNA. *Die Naturwissenschaften*, 1983; 70(5):252–4. doi:10.1007/BF00405447. PMID 6877387.

Kolaczkowska, E; Kubes, P. Neutrophil recruitment and function in health and inflammation. *Nat Rev Immunol,* 2013; 13:159-175.

Kolb, M; Bonniaud, P; Galt, T; Sime, P; Kelly, M; Margetts, P; Gauldie, J. Differences in the fibrogenic response after transfer of active transforming growth factor - β1 gene to lungs of “fibrosis-prone” and “fibrosis-resistant” mouse strains. *Am J Respir Cell Mol Biol,* 2002; 27:141-50.

Koning, RE. 1994. [Light.] *Plant Physiology Information Website*. Available at: http://plantphys.info/plant_physiology/light.shtml. (2-28-2016).

Kovalchuk, O; Baulch, JE. Epigenetic Changes and Nontargeted Radiation Effects - Is There a Link? *Environmental and MolecularMutagenesis*, 2008; 49:16-25.

Krasin, F; Wagner, H: Biological effects of nonionizing electromagnetic radiation. In: *Encyclopedia of Medical Devices and Instrumentation,* 1988.

Kroemer, G; Jaattela, M. Lysosomes and autophagy in cell death control. *Nat Rev Cancer*, 2005; 5(11):886-97.

Kuttler, F; Mai, S. Formation of non-random extrachromosomal elements during development, differentiation and oncogenesis". *Seminars in Cancer Biology,* 2007; 17(1):56-64. doi:10.1016/j.semcancer.2006.10.007. PMID 17116402.

Kuwana, T; Mackey, MR; Perkins, G; Ellisman, MH; Latterich, M; Schneiter, R; Green, DR; Newmeyer, DD. Bid, Bax, and lipids cooperate to form supramolecular openings in the outer mitochondrial membrane. *Cell*, 2002; 111:331-42.

Lamp, EC; Drexler, HG. Biology of tartrate-resistant acid phosphatase. *Leuk Lymphoma*, 2000; 39(5-6):477-84.

Lau, KH; Baylink, DJ. Osteoblastic tartrate-resistant acid phosphatase: its potential role in the molecular mechanism of osteogenic action of fluoride. *J Bone Miner Res*, 2003; 18(10):1897-900.

Lefeuvre, G; Gorodetzky, P; Dolbeau, J; Patzak, T; Salin, P. Absolute measurement of the nitrogen fluorescence yield in air between 300 and 430 nm. *Nucl Instrum Methods Phys Res*, 2007; A578:78–87.

Lehnert, BE; Iyer, R. Exposure to low-level chemicals and ionizing radiation: reactive oxygen species and cellular pathways. *Hum Exp Toxicol,* 2002; 21(2):65-9.

Leeper-Woodford, SK; Adkison, LR. The cellular and molecular basis for human systems. In: *Integrated Systems*. Leeper-Woodford, SK; Adkison, LR. (ed.). Wolters Kluwer, 2016. 1-38.

Levine, PH; Young, HA; Simmens, SJ; Rentz, D; Kofie, VE; Mahan, CM; Kang, HK. Is Testicular Cancer Related to Gulf War Deployment? Evidence from a Pilot Population-Based Study of Gulf War Era Veterans and Cancer Registries. *Military Medicine*, 2005; 170(2):149-153.

Li, B; Mahan, CM; Kang, HK; Eisen, SA; Engel, CC. Longitudinal Health Study of US 1991 Gulf War Veterans: Changes in Health Status at 10-Year Follow-up. *Am J Epidemiol,* 2011; 174(7):761–8. doi: 10.1093/aje/kwr154.

Limbach, LK; Wick, P; Manser, P; Grass, RN; Bruinink, A; Stark, WJ. Exposure of engineered nanoparticles to human lung epithelial cells: The influence of chemical composition and catalytic activity on oxidative stress. *Environ Sci Technol,* 2007; 41(11):4158-63. DOI: 10.1021 /es062629t.

Lin, MT; Beal, MF. Mitochondrial dysfunction and oxidative stress in neurodegenerative diseases. *Nature,* 2006; 443:787–95. doi:10.1038 /nature05292.

Lindsey, R. [Climate and Earth's Energy Budget]. [2009 January 14]. Available at: http://earthobservatory.nasa.gov/Features/EnergyBalance /page1.php. (2016 April 26).

Little, JB. Radiation carcinogenesis. *Carcinogenesis*, 2000; 21:397–404.

Lohman, BL; Welsh, RM. Apoptotic Regulation of T Cells and Absence of Immune Deficiency in Virus-Infected Gamma Interferon Receptor Knockout Mice. *J Virol,* 1998; 72(10):7815–21.

Lorimore, SA; Coates, PJ; Wright, EG. Radiation-induced genomic instability and bystander effects: interrelated nontargeted effects of exposure to ionizing radiation. *Oncogene*, 2003; 22(45):7058-69.

Lowe, XR; Bhattacharya, S; Marchetti, F; Wyrobek, AJ. Early brain response to low-dose radiation exposure involves molecular networks and pathways associated with cognitive functions, advanced aging and Alzheimer's disease. *Radiat Res*, 2009; 171(1):53-65. doi: 10.1667/RR1389.1.

Luchin, A; Purdom, G; Murphy, K; Clark, MY; Angel, N; Cassady, AI; Hume, DA; Ostrowski, MC. The Microphthalmia Transcription Factor Regulates Expression of the Tartrate-Resistant Acid Phosphatase Gene During Terminal Differentiation of Osteoclasts. *J Bone Miner Res*, 2000; 15:451-60.

Marinelli, F. [Cell Membrane and Electromagnetic Fields. Proceedings]. Ist. di Citomorfologia C.N.R., Bologna. [2007] Available at: https://www.salzburg.gv.at/gesundheit_/Documents/proceedings_%2813 %29_marinelli.pdf. (2016 April 26).

Markovic, J; Stojsic, J; Zunic, S; Ruzdijic, S; Tanic, N. Genomic instability in patients with non-small cell lung cancer assessed by the arbitrarily primed polymerase chain reaction. *Cancer investigation*, 2008; 26(3):262-8.

Martell, EA. Alpha-Radiation Dose at Bronchial Bifurcations of Smokers from Indoor Exposure to Radon Progeny. *Proc Nat Acad Sci USA*, 1983; 80(5):1285-9.

Mathews JD, Forsythe AV, Brady Z, Butler MW, Goergen SK, Byrnes GB, Giles GG, Wallace AB, Anderson PR, Guiver TA, McGale P. Cancer risk in 680 000 people exposed to computed tomography scans in childhood or adolescence: data linkage study of 11 million Australians. *BMJ,* 2013; 346:f2360. doi: http://dx.doi.org/10.1136/bmj.f2360.

Matsumoto, H; Takahashi, A; Ohnishi, T. Radiation-induced adaptive responses and bystander effects. *Biol Sci Space*, 2004; 18(4):247-54.

Menendez, JA; Lupu, R. Fatty acid synthase and the lipogenic phenotype in cancer pathogenesis. *Nature Reviews Cancer*, 2007; 7:763-77. doi:10.1038/nrc2222.

Mezentsev, A; Amundson, SA. Global gene expression responses to low-or high-dose radiaton in a human three-dimensional tssue model. *Radiaton Research*, 2011; 175(6):677-88.

Midthun, DE; Jett, JR; Screening for lung cancer. In: *Thoracic Malignancies. Eur Respir Mon.* Spiro, SG; Huber, RM; Janes, SM. (ed.). UK, 2009; 44:57–70. ISSN 1025-448x.

Mihailovic, J; Pechlivanoglou, P; Miladinov-Mikov, M; Živković, S; Postma, MJ. Cancer incidence and mortality in Serbia 1999–2009. *BMC Cancer*, 2013; 13:18. DOI**:** 10.1186/1471-2407-13-18.

Military Toxics Project Information Sheet (first version). ["Depleted" Uranium Munitions: Nuclear Waste as a Weapon]. [2003 June] Available at: http://www.dmzhawaii.org/wp-content/uploads/2009/02/depleted-uranium-fact-sheet.pdf.

Miller, AC; Xu, J; Stewart, M; Prasanna, PG; Page, N. Potential health effects of the heavy metals, depleted uranium and tungsten, used in armor-piercing munitions: comparison of neoplastic transformation, mutagenicity, genomic instability, and oncogenesis. *Met Ions Biol Med*, 2000; 6:209-11.

Miller, AC; Stewart, M; Brooks, K; Shi, L; Page, N. Depleted uranium-catalyzed oxidative DNA damage: absence of significant alpha particle decay. *J Inorg Biochem*, 2002; 91(1):246–52. doi:10.1016/S0162-0134(02)00391-4.

Miller, I. Schumann Resonance, Psychophysical Regulation & Psi (Part I). *Journal of Consciousness Exp loration & Research*, 2013; 4(6):599-612.

Milosevic, J; Rakic, Lj; Ruzdijic, S. Detection of apoptotic cells in the rat brain by TUNEL technique. *Arch Biol Sci*, 1998; 50(2):13P-14P.

Milosevic, J; Veskov, R; Vasilev, V; Rakic, Lj; & Ruzdijic, S. Apoptosis induction by phencyclidine in the brain of rats of different ages. *Addict Biol*, 2000;5:157-65.

Minic, N; Zunic, S; Zunic, S. NeuroStation–Statistical software based on artificial intelligence and pattern recognition for NSCLC development prediction through comprehensive biomarker analysis. *European Respiratory Journal*, 2011; 38(Suppl55):p4437.

MIT, 2011. MIT Technology Review. [Atmosphere Above Japan Heated Rapidly Before M9 Earthquake. Emerging Technology From the arXiv]. [2011 May 18]. Available at: http://www.technologyreview.com/view/424033/atmosphere-above-japan-heated-rapidly-before-m9-earthquake/.

Mitchel, REJ. Low doses of radiation are protective in vitro and in vivo: Evolutionary origins. *Dose-response*, 2006; 4(2):75-90.

Miyashita, T; Reed, JC. Tumor suppressor p53 is a direct transcriptional activator of human box gene. *Cell*, 1995; 80:293-9. Available at: http://drrajivdesaimd.com/?p=5498.

Mizushima, N. Autophagy: process and function. *Genes & Dev*, 2007; 21:2861-73. doi: 10.1101/gad.1599207.

Monahan, RA; Dvorak, HF; Dvorak, AM. Ultrastructural localization of nonspecific esterase activity in guinea pig and human monocytes, macrophages, and lymphocytes. *Blood*, 1981; 58(6):1089-99.

Monleau, M; De Meo, M; Paquet, F; Chazel, V; Dumenil, G; Donnadieu-Claraz, M. Genotoxic and Inflammatory Effects of Depleted Uranium Particles Inhaled by Rats. *Toxicol Sci*, 2006; 89(1):287–95. doi: 10.1093/toxsci/kfj010.

Moret, L. [U.S. Weapons Poison Europe. Radiation From Iraq War Detected in UK Atmosphere]. AmericanFreePres.net. [Issue #10/11. 2006, March 6th and 13th]. Available at: www.americanfreepress.net/html/weapons_poison_europe.html. (2016 April 26).

Morgan, WF. Non-targeted and delayed effects of exposure to ionizing radiation: I. Radiation-induced genomic instability and bystander effects in vitro. *Radiat Res*, 2003; 159(5):567–80. doi:10.1667/0033-7587(2003)159[0581:NADEOE]2.0.CO;2

Morgan, WF. 2003-1. Non-targeted and Delayed Effects of Exposure to Ionizing Radiation: II. Radiation-Induced Genomic Instability and Bystander Effects *In Vivo*, Clastogenic Factors and Transgenerational Effects. *Radiat Res*, 2003; 159:581–96.

Morgan WF. 2003-2. Non-targeted and delayed effects of exposure to ionizing radiation. II. Radiation-induced genomic instability and bystander effects in vivo, clastogenic factors and transgenerational effects. *Radiat Res*, 2003; 159:581-96.

Morgan, WF. 2003-3. Is there a common mechanism underlying genomic instability, bystander effects and other nontargeted effects of exposure to ionizing radiation? *Oncogene,* 2003; 22:7094-9.

Morgan, WF; Sowa, MB. Non-targeted bystander effects induced by ionizing radiation. *Mutat Res*, 2007; 616(1-2):159-64. Epub 2006 Nov 28.

Moss, DW. Changes in enzyme expression related to differentiation and regulatory factors: the acid phosphatase of osteoclasts and other macrophages. *Clin Chim Acta*, 1992; 209:131-8.

Mothersill, C; Seymour, C. Radiation-induced bystander and other non-targeted effects: novel intervention points in cancer therapy? *Curr Cancer Drug Targets*, 2006; 6(5):447-54.

Muggenburg, BA; Guilmette, RA; Hahn, FF; Diel, JH; Mauderly, JL; Seilkop, SK; Bruce, B; Boecker, BB. Radiotoxicity of Inhaled $^{239}PuO_2$ in Dogs. *Radiation Res,* 2008; 170(6):736-57. doi: http://dx.doi.org/ 10.1667/ RR1409.1.

Munroe, G. [Health Effects of DU - International Coalition to Ban Uranium Weapons]. Compiled on behalf of ICBUW. Grassroots Action For Peace. Second Edition [2004 October]. Available at: www.bandepleteduranium .org/en/docs/8.pdf. (2016 April 25)

Nagasawa, H; Little, JB. Induction of sister chromatid exchanges by extremely low doses of alpha-particles. *Cancer Research,* 1992; 52(22):6394–6. Available at: http://cancerres.aacrjournals.org/content/52/22/ 6394. full.pdf.

Nakamura, Y; Romberg, DJ; Tate, L. Ertl, RF; Kawamoto, M; Adachi, Y; Mio, T; Sisson, JH; Spurzem, JR; Rennard, SI. Cigarette smoke inhibits lung fibroblast proliferation and chemotaxis. *Am J Respir Crit Care Med*, 1995; 15:1497-503.

Nankova, BB; Chua, J; Mishra, R; Kobasiuk, CD; La Gamma, EF. Nicotinic Induction of Preproenkephalin and Tyrosine Hydroxylase Gene Expression in Butyrate-Differentiated Rat PC12 Cells: A Model for Adaptation to Gut-Derived Environmental Signals. *Pediatr Res*, 2003; 53:113-118.

NASA, 2003. [Dust and Fire in Southern Iraq]. [2003 March 21]. Available at: http://www. nasa.gov/centers/goddard/news/topstory/2003/0321iraq.html#23. (April 23rd 2016).

NASA, 2005. [Flashes in the Sky: Earth's Gamma-Ray Bursts Triggered by Lightning]. [2005 February 17] Available at: http://www.nasa.gov/vision /universe/solarsystem/rhessi_tgf.html#.VcMeG7WDrIg.

http://www.nasa.gov/vision/universe/solarsystem/rhessi_tgf.html#.VxwBjzE0aSp (2016 April 24).

NASA, 2009. Climate Forcings and Global Warming". January 14, 2009. Available at: http://earthobservatory.nasa.gov/Features/EnergyBalance/page7.php.

NASA, 2013. The Electromagnetic Spectrum. Updated: March 2013. Available at: http://imagine.gsfc.nasa.gov/science/toolbox/emspectrum1.html. (2016 April 26).

NASA, 2013-1. Regions of the Electromagnetic Spectrum. Last Updated: 05-Nov-2013. Available at: http://imagine.gsfc.nasa.gov/science/ toolbox/spectrum_chart.html.

NIOSH, 2011. NIOSH (National Institute for Occupational Safety and Health). [Occupational exposure to titanium dioxide]. Current Intelligence Bulletin 63: Publication No 2011-160. [2011, April]. Available at: http://www.cdc.gov/niosh/docs/2011-160/pdfs/2011-160.pdf.

Odoardi, F; Sie, C; Streyl, K; Ulaganathan, VK; Schläger, C; Lodygin, D; Heckelsmiller, K; Nietfeld, W; Ellwart, J; Klinkert, WE; Lottaz, C; Nosov; M; Brinkmann, V; Spang, R; Lehrach, H; Vingron, M; Wekerle, H; Flugel-Koch, C; Flugel, A. T cells become licensed in the lung to enter the central nervous system. *Nature*, 2012; 488(7413):675-9.

Odum, EP; Barrett, GW. The Scope of Ecology. In: Fundamentals of ecology. Vol. 3. Odum, EP; Odum HT; Andrews J. (ed.). Philadelphia: Saunders, 1971. Available at: http://www.evtfuto.com/download/Other%20Links/FUNDAMENTALS%20OF%20ECOLOGY-P.Odum.pdf1.

Oleinick, NL; Chiu, SM; Friedman, LR. Gamma radiation as a probe of chromatin structure: Damage to and repair of active chromatin in the metaphase chromosome. *Radiat Res*, 1984; 98(3):629–41.

Oritz, LA; Moroz, K; Liu, JY; Hoyle, GW; Hammond, T; Hamilton, RF; Holian, A; Banks, W; Brody, AR; Friedman, M. Alveolar macrophage apoptosis and TNF-alpha, but not p53, expression correlate with murine response to bleomycin. *Am J Physiol*, 1998; 275:L1208-18.

Ortega, R; Bresson, C; Darolles, C; Gautier, C; Roudeau, S; Perrin, L; Janin, M; Floriani, M; Aloin, V; Carmona, A; Malard, V. Low-solubility particles and a Trojan-horse type mechanism of toxicity: the case of cobalt oxide on human lung cells. *Part Fibre Toxicol*, 2014; 11:14. doi: 10.1186/1743-8977-11-14.

Osman, M; Cantor, JO; Roffman, S; Keller, S; Turino, GM; Mandl, I. Cigarette smoke impairs elastin resynthesis in lungs of hamsters with elastase-induced emphysema. *Am Rev Respir Dis*, 1985; 132:640-3.

Ottaviani, E; Franceschi, F. The invertebrate phagocytic immunocyte: clues to a common evolution of immune and neuroendocrine systems. *Immunol Today*, 1997; 18:169-74.

Oyama, K. [Alpha radiation detected near Japan's west coast in Niigata, Yamagata — Local official suspects levels too high to be from Fukushima Daiichi (VIDEO)] by Fukushima Diary [2012 May 9]. Available at: http://enenews.com/alpha-radiation-detected-near-japans-west-coast-in-niigata-yamagata-local-official-suspects-levels-too-high-to-be-from-fukushima-daiichi-video.

Park, WY; Hwang, CI; Im, C N; Kang, MJ; Woo, JH; Kim, JH; Kim, YS; Kim, JH; Kim, H; Kim, KA; Yu, HJ; Lee, SJ; Seo, JS. Identification of radiation-specific responses from gene expression profile. *Oncogene*, 2002; 21(55):8521-8.

Park, EJ; Yi, J; Kim, Y; Choi, K; Park, K. Silver nanoparticles induce cytotoxicity by a Trojan-horse type mechanism. *Toxicology in Vitro,* 2010; 24(3):872-878. doi:10.1016/j.tiv.2009.12.001.

Perkins, S. Dust, the thermostat: How tiny airborne particles manipulate global climate. *Sci News*, 2001; 160(13):200-201. DOI: 10.2307/4012776.

Pesnya, DS; Romanovsky, AV. Comparison of cytotoxic and genotoxic effects of plutonium-239 alpha particles and mobile phone GSM 900 radiation in the Allium cepa test. *Mutat Res*, 2013; 750(1-2):27-33. doi: 10.1016/j.mrgentox.2012.08.010.

Petkau, A. Effect of $22Na^{+}$ on a phospholipid membrane. *Health Physics,* 1972; 22(3):239. doi:10.1097/00004032-197203000-00004.

PHE, 2007. [Depleted uranium (DU): general information and toxicology. Information on depleted uranium (DU) including what it is, its uses, health effects and protection from exposure]. Part of: Radiation, Health protection. Public Health England) [2007 December 17]. Available at: https://www.gov.uk/depleted-uranium-du-general-information-and-toxicology.

Plowman, P; Flemans, R. Human pulmonary macrophages: The relationship of smoking to the presence of sea blue granules and surfactant turnover. *J Clin Path*ol, 1981; 33(8):738-43.

Pollycove, M. Nonlinearity of radiation health effects. *Environ Health Perspect*, 1998; 106(Suppl 1):363-8.

Pollycove, M; Paperiello, CJ. Health effects of low-dose radiation: Molecular, cellular, and biosystem response. In: *Low doses of ionizing radiation. Biological effects and regulatory control*. International Atomic Energy

Agency. Vienna: IAEA: IAEA-TECDOC-976 (IAEA-CN-67/63). 1997; 223-6.

Popovic, D; Todorovic, D; Frontasyeva, M; Ajtic, J; Tasic, M; Rajsic, S. Radionuclides and heavy metals in Borovac, Southern Serbia. *Environmental Science and Pollution Research*, 2008; 15(6):509-20.

Pouget, JP; Santoro, L; Raymond, L; Chouin, N; Bardiès, M; Bascoul-Mollevi, C; Huguet, H; Azria, D; Kotzki, PO; Pèlegrin, M; Vivès, E; Pèlegrin, A. Cell Membrane is a More Sensitive Target than Cytoplasm to Dense Ionization Produced by Auger Electrons. *Radiation Research*, 2008; 170(2): 192-200.

Pouget, JP; Lozza, C; Deshayes, E; Boudousq, V; Navarro-Teulon, I. Introduction to radiobiology of targeted radionuclide therapy. *Front Med (Lausanne),* 2015; 2:12. doi: 10.3389/fmed.2015.00012. eCollection 2015.

Preston, RJ; Boice, JD; Brill, AB; Chakraborty, R; Conolly, R; Hoffman, FO; Hornung, RW; Kocher, DC; Land, CE; Shore, RE; Woloschak, GE. Uncertainties in estimating health risks associated with exposure to ionising radiation. *J Radiol Prot,* 2013; 33(3):573–88. doi:10.1088/0952-4746/33/3/573.

Priest, ND. The distribution and behaviour of metals in the skeleton and body: studies with bone seeking radionuclides. In: *Trace metals and fluoride in bones and teeth.* Eds. Priest, ND; Van de Vyver, F. Boca Raton: CRC Press, 1990: 83–140.

Priest, ND. Toxicity of depleted uranium. *The Lancet*, 2001; 357:244-6. DOI: http://dx.doi.org/10.1016/S0140-6736(00)03605-9.

Prise, KM; Belyakov, OV; Newman, HC; Patel, S; Schettino, G; Folkard, M. Non-targeted Effects of Radiation: Bystander Responses in Cell and Tissue Models. *Radiat Prot Dosimetry,* 2002; 99(1-4):223-6.

Puchelle, E; Zahm, JM. Repair processes of the airway epithelium. In: *Airways and Environments: From injury to repair; Lung biology in Health and Disease*. Lenfant, CL. (ed.). Marcel Dekker, New York, 1996; 157-82.

Puchelle, E. Repair and regeneration of the airway Epithelium after injury. In : Current perspectives in epithelial cell injury and repair: consequences for epithelial functions. *ERS Research Seminars*, 2003; 9-11.

Radak, Dj. Immunological parametres in the experimental arteriosclerosis. *Master thesis.* Belgrade University, Medical Faculty, 1983.

Ragheb, M. [Gamma rays interaction with matter]. [2013 October 30]. Available at: http://mragheb.com/NPRE%20402%20ME%20405%20 Nuclear%20Power%20Engineering/Gamma%20Rays%20Interactions%20 with%20Matter.pdf.

Rajiv, D. [Cell Death]. [2014 January 1st] 2014. Available at: http://drrajivdesaimd. com/?p=5498.

Randers-Pehrson, G; Geard, CR; Johnson, G; Ellistion, CD; Brenner, DJ. The Columbia University single-ion microbeam. *Radiat Res*, 2001; 156:210-14.

Rekacewicz, P. Forest fires in Albania, Macedonia, Montenegro, Serbia and Bulgaria. [Balkan Vital Graphics]. *Environment without borders UNEP/GRID-Arendal* [2007]. Available at: http://www.grida.no/files/publications/balkan-vital-graphics/balkans-vital-graphic-full.pdf.

Reshetnyak, SA; Shcheglov, VA; Blagodatskikh, VI; Gariaev, PP; Maslov, MY. Mechanisms of interaction of electromagnetic radiation with a biosystem. *Laser Physics-Lawrence*. 1996; 6:621-53.

Reynaud, E. Protein Misfolding and Degenerative Diseases. *Nature*, 2010; 3(9):28.

Reynolds, H. Bronchoalveolar lavage. *Am Rev Respir Dis*, 1987; 135:250-63.

Rouet-Benzineb, P; Aparicio, T; Guilmeau, S; Pouzet, C; Descatoire, V; Buyse, M; Bado, A. Leptin Counteracts Sodium Butyrate-induced Apoptosis in Human Colon Cancer HT-29 Cells via NF-κB Signaling. *J Biol Chem*, 2004; 279(16):16495-502.

Roy, L; Gruel, G; Vaurijoux, A. Cell response to ionizing radiation analyzed by gene expression patterns. *Ann Ist Super Sanità*, 2009; 45(3):272 -7.

RS, 2001. The health hazards of depleted uranium munitions-Part 1. London: Royal Society, 2001.

RS, 2002. The health hazards of depleted uranium munitions Part II. *Policy Document* 5/02. London: Royal Society, 2002.

Rush, M; Misra, R. Extrachromosomal DNA in eukaryotes. *Plasmid*, 1985; 14(3):177–91. doi:10.1016/0147-619X(85)90001-0. PMID 3912782.

Sabella. S; Carney, RP; Brunetti, V; Malvindi, MA; Al-Juffali, N; Vecchio, G; Janes, SM; Bakr, OM; Cingolani, R; Stellacci, F; Pompa, PP. A general mechanism for intracellular toxicity of metal-containing nanoparticles. *Nanoscale*, 2014; 21;6(12):7052-61. doi: 10.1039/c4nr01234h.

Sand, J; Ihantola, S; Peräjärvi, K; Toivonen, H; Toivonen, J. Radioluminescence yield of alpha particles in air. *New Journal of Physics*, 2014; 16:053022. doi:10.1088/1367-2630/16/5/053022.

Saunders, RD; Jefferys, JG. Weak electric field interactionsin the central nervous system. *Health Phys*, 2002; 83(3):366-75.

Sawant, SG; Randers-Pehrson, G; Geard, CR; Brenner, DJ; Hall EJ. Radiat Res, 2001; 155:397-401.

Scaffidi, C; Kirchhoff, S; Krammer, PH; Peter, ME. Apoptosis signaling in lymphocytes. *Curr Opin Immunol*, 1999; 11:277-85.

Schaberg, T. Smoking and the host defense system: Effects on physical defense mechanisms and immune responses. In: ERS, eds. Smoking and respiratory Tract. World Congress on Lung Health and 10th ERS Annual Congress: Florence, 2000; pp: 4-5.

Schaefer, DJ; Bourland, JD; Nyenhuis, JA. Review of patient safety in time-varying gradient fields. *J Magn Reson Imaging*, 2000; 12:20-29.

Schwarz WBK. [Depleted Uranium In India, Spreading Worldwide]. [2006 March 20]. Available at: http://www.rense.com/general70/deple.htm.

Schwartz, LM; Osborne, BA. Programmed cell death, apoptosis and killer genes. *Immunol Today*, 1993; 14:582-90.

Scott, BR. Stochastic thresholds: A novel explanation of nonlinear dose-response relationships. *Dose-Response*, 2005; 3:547-67.

Scott, BR. Low-dose-radiation stimulated natural chemical and biological protection against lung cancer. *Dose Response,* 2008; 6(3):299-318. doi: 10.2203/dose-response.07-025.Scott.

Scott, BR; Di Palma, J. Sparsely ionizing diagnostic natural background radiation are likely preventing cancer and other genomic-instability-associated diseases. *Dose-Response*, 2007; 5(3): 230-55.

Seideman, JH; Stancevic, B; Rotoloa, JA; McDevittc, MR; Howelld, RW; Kolesnicka, RN; Scheinberga, DA. Alpha Particles Induce Apoptosis through the Sphingomyelin Pathway. *Radiat Res*, 2011; 176(4):434-6.

Semikhina, LP; Kiselev, VF. Effect of weak magnetic fields on the properties of water and ice. *Russian Physics Journal*, 1981; 31:5351-4.

Semikhina, LP; Kiselev, VF; Levshin, LV; Saletskii, AM. Effect of weak magnetic fields on the luminescence - spectral properties of a dye in an aqueous solution. *Journal of Applied Spectroscopy*, 1988; 48:556-9.

Shields, PG; Harris, CC. Cancer risk and low-penetrance susceptibility genes in gene-environment interactions. *J Clin Oncol*, 2000, 18 (11):2309-15.

Shikazono, N; Noguchi, M; Fujii, K; Urushibara, A; Yokoya, A. The yield, processing, and biological consequences of clustered DNA damage induced by ionizing radiation. *J Radiat Res*, 2009; 50(1):27-36.

Shoji, S; Ertl, RF; Linder, J. Romberger, DJ; Rennard, SI. Bronchial epithelial cells produce chemotactic activity for bronchial epithelial cells: possible role for fibronectin in airway repair. *Am Rev Respir Dis,* 1990; 141:218-25.

Shuryak, I; Sachs, RK; Brenner, DJ. Biophysical Models of Radiation Bystander Effects: 1. Spatial Effects in Three-Dimensional Tissues. *Radiat Res,* 2007; 168:741–9. doi: http://dx.doi.org/10.1667/RR1117.1.

Siafakis, NM; Tzortzaki, EG. Few smokers develop COPD. Why? *Respir Med,* 2002; 96: 615-24.

Sienkiewicz, Z. Rapporteur report: Other tissues. *Radiat Prot Dosimetry*, 2003; 106(4):391-6.

Simões, F; Pfaff, R; Freudenreich, H. Satellite observations of Schumann resonances in the Earth's ionosphere. *Geophysical Research Letters*, 2011; 38(22):L22101. doi:10.1029/2011GL049668

Smith, B. Introduction to radioactivity and radioactive decay. In: *Nuclear Pharmacy. Concepts and applications in nuclear pharmacy*. Pharmaceutical Press. 2010; pp. 13-53.

Smith, RA; Manassaram-Baptiste, D; Brooks, D; Doroshenk, M; Fedewa, S; Saslow, D; Brawley, OW; Wender, R. Cancer Screening in the United States, 2015: A Review of Current American Cancer Society Guidelines and Current Issues in Cancer Screening. *CA Cancer J Clin*, 2015; 65(1):30-54. doi: 10.3322/caac.21261.

Sorokin, V; Hayakawa M. Generation of Seismic-Related DC Electric Fields and Lithosphere-Atmosphere-Ionosphere Coupling. *Mod Appl Sci*, 2013; 7(6):1-25. doi:10.5539/mas.v7n6p1.

Sousa, N. The dynamics of the stress neuromatrix. *Molecular Psychiatry*, 2016; 21:302-12; doi:10.1038/mp.2015.196.

Stein, RS; Toda, S. Megacity Megaquakes-Two Near Misses. *Science,* 2013; 341(6148):850-2. doi: 10.1126/science.1238944.

Stuart, L; Hughes, J. Apoptosis and autoimmunity. *Nephrol Dial Transplant*, 2002; 17(5):697-700. doi: 10.1093/ndt/17.5.697.

Swarts, SG; Gilbert, DC; Sharma, KK; Razskazovskiy, Y; Purkayastha, S; Naumenko, KA; Bernhard, WA. Mechanisms of Direct Radiation Damage in DNA, Based on a Study of the Yields of Base Damage, Deoxyribose Damage, and Trapped Radicals in $d(GCACGCGTGC)_2$. *Radiat Res*, 2007; 68(3):367-81. doi: 10.1667/RR1058.1.

Szumiel, I. Radiation hormesis: Autophagy and other cellular mechanisms. *Int J Radiat Biol*, 2012; 88(9):619-28. doi: 10.3109/09553002.2012.699698.

Tajmir-Riahi, HA. D-glucose interaction with uranium ion. Synthesis, spectroscopic and structural characterization of uranyl-glucose adducts and the effect of metal cation binding on the sugar anomeric structures. *Inorganica chimica acta*, 1988; 153(3):155-9.

Tartier, L; Gilchrist, S; Burdak-Rothkamm, S; Folkard, M; Prise, KM; Cytoplasmic irradiation induces mitochondrial-dependent 53BP1 protein relocalization in irradiated and bystander cells. *Cancer Res,* 2007; 67:5872–5879. doi: 10.1158/0008-5472.CAN-07-0188.

Tasat, DR.; Orona, NS; Bozal, C; Ubios, AM; Cabrini, RL. Intercellular Metabolism of Uranium and the Effects of Bisphosphonates on Its Toxicity. In *Cell Metabolism - Cell Homeostasis and Stress Response.* Bubulya, P. (ed.). Tech Publishers, 2012.

Tipler, PA; Llewellyn, R. Modern Physics (4th ed.). Freeman, WH. 2002. ISBN 0-7167-4345-0.

Toma, CL; Serbescu, A; Alexe, M; Cervis, L; Ionita, D; Bogdan, MA. The Bronchoalveolar Lavage Pattern in Radiation Pneumonitis Secondary to Radiotherapy for Breast Cancer. *Mædica*. 2010; 5(4):250-257.

Tubiana, M. Prevention of cancer and the dose–effect relationship: The carcinogenic effects of ionizing radiations. *Cancer/Radiothérapie*, 2009; 13:238-58. doi:10.1016/j.canrad.2009.03.003.

Tubiana, M; Feinendegen, LE; Yang, C; Kaminski, JM. The Linear No-Threshold Relationship is inconsistent with radiation biologic and experimental data. *Radiology*, 2009; 251(1):13-22, doi:10.1148/radiol.2511080671.

Ulrich, A; Mayer, A; Kasper, M; Wichser, A; Czerwiński, J. Emission of metal-oxide particles from IC-engines. *Silniki Spalinowe*, 2011; 50:72-88.

Ulsh, BA. The New Radiobiology: Returning to Our Roots. *Dose-Response*, 2012; 10(4):593-609. doi:10.2203/dose-response.12-021.Ulsh.

UNSCEAR, 2000. United Nations Scientific Committee on the Effects of Atomic Radiation, UNSECAR 2000 report. *Volume II: Effects. Effects of Atomic Radiation*, 2000. Available at: http://www.unscear.org/docs/reports/annexg.pdf.

UNSECAR, 2006. United Nations Scientific Committee on the Effects of Atomic Radiation, UNSCEAR 2006 Report to the General Assembly with Scientific Annexes. Sources and Effects of Ionizing Radiation. *Volume I Report and Annexes A and B.* New York, United Nations, 2008.

U.S. Atomic Energy Commission, 1974. Regulatory Guide 5.21. [Nondestructive Uranium-235 enrichment assay by gamma-ray spectrometry]. Directorate of regulatory standards, 1974: 1-7. (April 22nd 2016). Available at: http://www.nrc.gov/docs/ML1306/ML13064A082.pdf.

Vaiserman, AM. Radiation Hormesis: Historical Perspective and Implications for Low-Dose Cancer Risk Assessment. *Dose-Response.* 2010; 8(2):172-191. doi:10.2203/dose-response.09-037. Vaiserman.

Vincent, JB; Averill, BA. An enzyme with a double identity: purple acid phosphatase and tartrate-resistant acid phosphatase. *FASEB J,* 1990; 4(12):3009-14.

Waldenmaier, T. Spectral resolved measurement of the nitrogen fluorescence yield in air induced by electrons. 2006. Available at: http://bibliothek.fzk.de/zb/berichte/FZKA7209.pdf.

Westermark, P; Andersson, A; Westermark. GT. Islet Amyloid Polypeptide, Islet Amyloid, and Diabetes Mellitus. *Physiological Reviews,* 2011; 91(3):795-826; DOI: 10.1152/physrev.00042.2009.

WHO, 2001. [Health effects of depleted uranium] A54/19. World Health Organization. Fifty-fourth World Health Assembly. [2001 April 26]. Available at: www.who.int/environmental_information/radiation/depleted_uranium.htm.

WHO, 2001-1. [Depleted uranium Sources, Exposure and Health Effects. WHO/SDE/PHE/01.1] World Health Organization Geneva. Department of Protection of the Human Environment. [2001 April]. Available at: http://whqlibdoc.who.int/hq/2001/who_sde_phe_01.1.pdf.

WHO, 2003. [Depleted Uranium. Fact Sheet N°257]. World Health Organization. [2003 April 10]. Available at: http://www.navend.de/html/aktuell/irak-krise/DU/WHO-factsheet257.pdf.

WHO, 2007. World Health Organization Environmental Health Criteria 238 – Extremely Low Frequency Fields. World Health Organization. 2007. Available at: http://docs.stetzerelectric.com/WHO-extremely-low-frequency-fields.pdf

Widel, M; Przybyszewski, W; Rzeszowska-Wolny, J. Radiation-induced bystander effect: The important part of ionizing radiation response. Potential clinical implications. *Postepy Hig Med Dosw,* 2009; 63:377-88.

WISE, 2011. [Uranium Project 2011]. World Information Service of Energy. http://www.wise-uranium.org/rup.html.

WPL, 2016. Extrachromosomal DNA in Eukaryotes. Extrachromosomal circular DNA. World Public Library, 2016. Available at: http://www.worldlibrary.org/articles/extrachromosomal_dna#Extrachromosomal_DNA_in_Eukaryotes.

Wright, EG. Commentary on radiation-induced bystander effects. *Hum Exp Toxicol,* 2004; 23(2):91-4.

WSDH, 2002. [Background Radiation Natural versus Man-Made. Washington State Department of Health]. Division of Environmental Health Office of Radiation Protection. Fact Sheet 320-063. [2002 July]. Available at: http://www.doh.wa.gov/portals/1/Documents/Pubs/320-063_bkvsman_fs.pdf. (2016 April 26).

Wyrobek, AJ; Manohar, CF; Krishnan, VV; Nelson, DO; Furtado, MR; Bhattacharya, MS; Marchetti, F; Coleman, MA. Low dose radiation response curves, networks and pathways in human lymphoblastoid cells exposed from 1 to 10cGy of acute gamma radiation. *Mutat Res*, 2011; 17;722(2):119-30. doi: 10.1016/j.mrgentox.2011.03.002.

Xia, T; Kovochich, M; Brant, J; Hotze, M; Sempf, J; Oberley, T; Sioutas, C; Yeh, JI; Wiesner, MR; Nel, AE. Comparison of the abilities of ambient and manufactured nanoparticles to induce cellular toxicity according to an oxidative stress paradigm. *Nano Lett,* 2006; 6(8):1794-807.

Yin, E; Nelson, DO; Coleman, MA; Peterson, LE; Wyrobek, AJ. Gene expression changes in mouse brain after exposure to low-dose ionizing radiation. *Int J Radiat Biol,* 2003; 79, 759-75.

Yuan, X-M; Li, W; Dalen, H; Lotem, J; Kama, R; Sachs, L; Brunk, UT. Lysosomal destabilization in p53-induced apoptosis. *Proceedings of the National Academy of Sciences of the United States of America*, 2002; 99(9):6286-91. doi:10.1073/pnas.092135599.

Zheng, MH; Lau, TT; Prince, R; Criddle, A; Wysocki, S; Beilharz, M; Papadimitriou, JM; Wood, DJ. 17beta-estradiol suppresses gene expression of tartrate-resistant acid phosphatase and carbonic anhydrase II in ovariectomized rats. *Calcif Tissue Int*, 1995; 56 (2):166-9.

Zinets, N; Prentice, A. [Chernobyl: Huge forest fire risking radioactive spread now poses 'no risk' to abandoned nuclear power plant]. Mirror [2015 Apr 29, 20:04]. Available at: http://www.mirror.co.uk/news/world-news/chernobyl-huge-forest-fire-risking-5600635.

Zunic, S. Cytochemical parameters of alveolar macrophages in sarcoidosis and malignant lung diseases. *Doctoral Thesis*. Belgrade University, Medical Faculty, 1993.

Zunic, S. Cytological characteristics of lung washings from children in depleted uranium stroked region. *J Biol Regul Homeost Agents,* 2013; 27(4):961-7.

Zunic, S. 2013-1. Lupus erythematosus cell phenomenon in pediatric bronchoalveolar lavages: possible manifestation of early radioadaptive response in radiation induced alveolitis. *J Biol Regul Homeost Agents.* 2013-1; 27(2):389-98.

Zunic, S. Hormesis-threshold model for individualized lung cancer risk assessment in smoking and low-dose radiation exposure. *Prague Medical Report*, 2015; 116 (Supplement): 43-4.

Zunic, S. 2016. Odabrana poglavlja iz medicinske biohemije. Uvod u energetski metabolizam. Medija centar "Odbrana", Beograd, 2016. *in press*.

Zunic, S; Minic, P; Kokai, G; Minic, A. Is the Lupus Erythematosus Cell Phenomenon detected in bronchoalveolar lavage samples a nonspecific insight into pathogenesis of different pulmonary diseases? *Immunology & Infectious Diseases*, 1996; 6:203-5.

Zunic, S; Milosevic, J; Ruzdijic, S; Zunic, S; Djordjevic-Denic, G; Rakic, Lj. Semiquantitative cytochemistry in evaluation of apoptotic capacity in broncholaveolar lavage of smokers and patients with non-small-cell-lung cancer. *J Biol Regul Homeost Agents*, 2004; 18(3/4):372-80.

Zunic, S; Rakic, L; Zunic-Bozinovski, S; Minic, N; Cytochemical and Apoptotic Markers of Tissue Remodeling, In: *Frontiers in Cell Apoptosis Research*. Erlich, SR. (ed). Nova Publishers: New York, 2007; pp 37 – 104.

Zunic, SS; Rakic, LM. Hormetic effects of depleted uranium to the biosphere and lithosphere-atmosphere-ionosphere coupling. *J Environ Occup Sci*, 2013; 2(2):103-7. doi:10.5455/jeos.20130911102811.

Zunic, S; Rakic, Lj. Environmental and health effects of depleted uranium. Chapter 3. In: *Uranium: Sources, Exposure and Environmental Effects*. Nelson, JR. (ed.). Nova Science Publishers. New York, 2015; pp. 53-86.

INDEX

α

A

B

C

D

E

F

G

H

I

K

L

M

N

O

P

R

S

T

U

V

W

X

Y